Advanced Intelligent Pipeline Management Technology

Huai Su · Qi Liao · Haoran Zhang · Enrico Zio
Editors

Advanced Intelligent Pipeline Management Technology

Springer

Editors
Huai Su
National Engineering Laboratory for Pipeline Safety/MOE Key Laboratory of Petroleum Engineering, College of Machinery and Transportation Engineering
China University of Petroleum
Beijing, China

Qi Liao
National Engineering Laboratory for Pipeline Safety/MOE Key Laboratory of Petroleum Engineering, College of Machinery and Transportation Engineering
China University of Petroleum
Beijing, China

Haoran Zhang
School of Urban Planning and Design
Peking University
Shenzhen, China

Enrico Zio
Centre for research on Risks and Crises (CRC)
Mines Paris-PSL University
Sophia Antipolis, France

Department of Energy
Politecnico di Milano
Milan, Italy

ISBN 978-981-19-9898-0 ISBN 978-981-19-9899-7 (eBook)
https://doi.org/10.1007/978-981-19-9899-7

This Springer imprint is published by the registered company Springer Nature Singapore Pte Ltd.
The registered company address is: 152 Beach Road, #21-01/04 Gateway East, Singapore 189721, Singapore

Acknowledgments

The editors would like to thank all of those colleagues from our own institution and those from further afield with whom we have talked about our ideas, our past and present students, and our friends and families.

In particular, we would like to extend our thanks to

PipeChina South Company for the advice and data support.
Professor Jinjun Zhang and *Prof. Yongtu Liang* for their support and encouragement in the early stage of this book.
Professor Enrico Zio, Associate Professor Huai Su, Associate Professor Haoran Zhang, and *Lecturer Qi Liao* for their editorial advice and support.

The editors would like to especially thank the funding supported by ***the National Natural Science Foundation of China (51904316)*** and ***the National Natural Science Foundation of China (51874325)***, and the research fund provided by ***the China University of Petroleum, Beijing (2462018YJRC038)***.

Contents

Chapter 1
Overview for Pipeline Scheduling

Renfu Tu and Zhengbing Li

Abstract Oil and gas pipe networks play an important role in ensuring the downstream energy supply as they transport about 60% of oil and gas. To guarantee the safe and stable operation of pipelines during the scheduling horizon, it is necessary to make a reasonable schedule. During the past two decades, the pipeline scheduling problem has been widely studied in the world. This chapter conducts quantitative analysis and visual research on related literature. Through the review of the existing articles, the research framework of pipeline scheduling problem is constructed, and their research defects and trends are also analyzed. The aim of this chapter is to provide a comprehensive understanding of pipeline scheduling as well as some directions and inspiration for the future research.

1.1 Pipeline Scheduling Content

Among multiple transportation modes, pipelines are the most effective mode for transporting large amounts of oil and gas resources over long distances because they have the advantages of low operating costs, high reliability, low environmental pollution, and less susceptibility to severe weather [1]. The advancement of pipeline scheduling research is closely related to the development of pipelines. The total length of pipelines in the world will continue to increase year by year, reaching a total of 201.9×10^4 km of oil and gas pipelines by the end of 2020. Among them, the top three countries are the USA, Russia, and China, with a total mileage of 72.47×10^4 km, 25.99×10^4 km, and 14.5×10^4 km, respectively. Liquid pipelines connect locations such as oil fields, crude oil depots, refineries, and product depots (as shown in Fig. 1.1), and gas pipelines connect locations such as gas fields, gas storage depots, and urban gate stations (as shown in Fig. 1.2). Overall, oil and gas

R. Tu (✉) · Z. Li
Beijing Key Laboratory of Urban Oil and Gas Distribution Technology, China University of Petroleum-Beijing, Fuxue Road No.18, Changping 102249, Beijing, China
e-mail: turenfu@foxmail.com

Z. Li
e-mail: cup_lzb@163.com

H. Su et al. (eds.), *Advanced Intelligent Pipeline Management Technology*,
https://doi.org/10.1007/978-981-19-9899-7_1

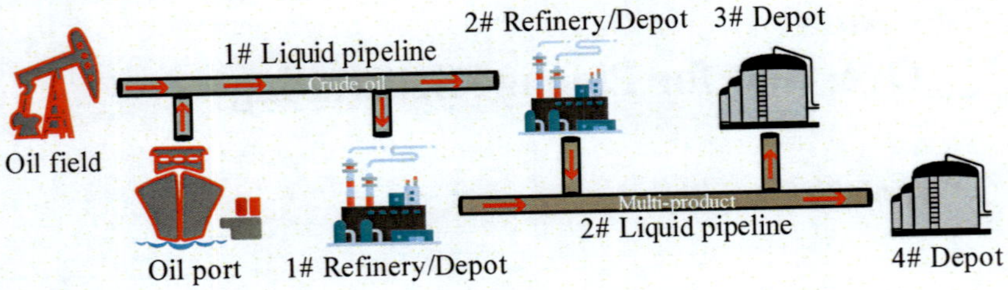

Fig. 1.1 Liquid piping system structure schematic diagram

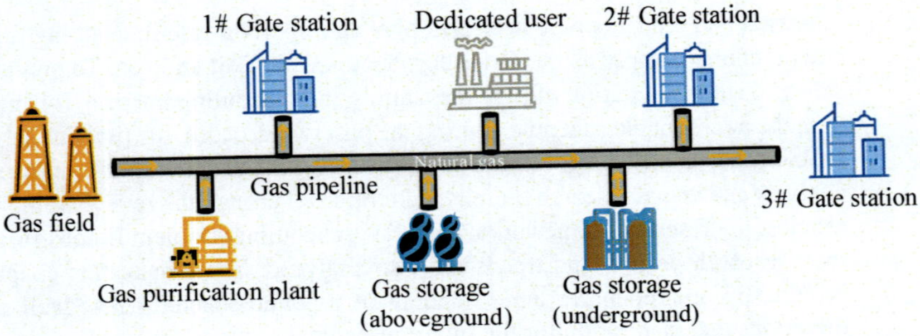

Fig. 1.2 Gas pipeline system structure schematic diagram

pipeline networks play an important role in securing downstream energy supplies, as they transport approximately 60% of oil and gas. Whether it is natural gas or crude oil and refined products, the different product properties, downstream demand, and other specificities make oil and gas pipelines usually transported in batches, and this is especially evident in multi-product pipelines [2, 3]. In the operation and management of oil and gas pipelines, the development of pipeline scheduling is critical, and the rationality of the schedule will affect the operational safety of the pipeline and the downstream supply stability. In production sites, schedulers mainly use manual methods to make plans, which requires a high level of professionalism and rich working experience of schedulers. Certainly, the scheduling of gas pipelines and liquid pipelines often differs greatly due to the different physical and chemical properties and uses of the media transported in the pipelines.

1.1.1 Gas Pipeline Network

According to the stage production characteristics, natural gas pipeline network scheduling needs to monitor and master the dynamic changes of upstream and downstream users at any time during the implementation process in accordance with the sales plan of the enterprise, and centrally control the key links in the gas transmission process, so that the upstream, middle, and downstream resources are steadily scheduling and connected in a consistent manner, thus satisfying the gas demand of each downstream user while ensuring the safe and stable operation of the pipeline. Therefore, natural gas scheduling must control all aspects of production operation in a comprehensive, dynamic, and timely manner and achieve the production purpose of unified organization and unified command by controlling and coordinating all aspects of production operation in a systematic, rational, and dynamic manner, so as to ensure the safe, efficient, and low-consumption completion of transmission tasks [4]. The gas pipeline operation optimization solves this problem by reasonably formulating the gas transmission plan and its operation optimization scheme (including optimal scheduling plan, determination of steady-state optimal conditions, optimal control of non-steady-state conditions, etc.) to meet the given conditions, such as the structure of the pipeline network, the equipment configuration and performance of the pressure stations, the supply conditions of each gas source, the demand gas quantity of each customer, etc., so as to achieve or approach the optimal criteria requirements. The optimality criterion is a criterion for evaluating the merits of the operation scheme, which is determined according to the purpose of operation optimization. In most cases, the optimal criterion is the lowest operational energy consumption of the system. Therefore, for the gas transmission system, it is to make it complete the given gas transmission task within the specified period while operating with the lowest energy consumption, reduce the mismatch at low transmission volumes, and reduce or even eliminate the throttling losses.

1.1.2 Oil Pipeline Network

The scheduling of crude oil pipelines usually starts from port and involves the stages of oil unloading, storage, transmission, and reprocessing, covering multiple ports, tank storage areas, pipelines, and refineries [5]. Meanwhile, considering the large differences in the nature of crude oil from different import sources, the scheduling problem of crude oil pipeline network makes reasonable arrangements for when and under what conditions to transport crude oil with different condensation points and how to coordinate the operation of pipelines within the network. In the crude oil pipeline network scheduling system, the scheduling of oil depot storage tanks cannot be ignored, and each depot distributes crude oil according to the refining demand of the downstream refineries. Given that the distance between each depot and the refinery and the transportation conditions in the area where the refinery is located vary, it makes the same refinery choose different sources of oil products. At the same time, the total distribution volume of a depot will be limited by its inventory and turnover factor, and it may happen that a certain depot cannot meet the oil supply demand of the neighboring refineries and thus needs oil from a distant depot to supplement.

As a bridge connecting upstream resources and downstream product consumption markets, multi-product pipelines are mainly composed of head stations, pump stations, delivery stations, injection stations, pipe sections, and other auxiliary related equipment, which are used to complete the injection, delivery, and transfer of multiple oil products. For multi-product pipelines, its scheduling plan is formulated to clarify the type, quantity, time window, and product availability of oil demand at each station and provide a basis for schedulers to prepare a reasonable scheduling plan. In the process of scheduling plan preparation, it is necessary to determine the operation management method, batch of products, batch order, start time and end time of delivery, flow rate value, pump start/stop scheme, and control and handling of oil mixing. Its main production equipment are pump and prime mover set and its auxiliary and control system, oil mixing detection device, pipe cleaner receiving and sending device, filtration device, metering calibration system, oil storage tank, pressure regulation, hydraulic drainage device, etc. At present, multi-product pipelines generally adopt normal temperature and closed transmission, when the pressure and flow rate in the pipeline change, it will have an impact on the operation of the whole pipeline system. Most multi-product pipelines have multiple branches and multiple outlets to meet the demand for oil delivery to cities along and near the pipeline route. Some pipelines may also have multiple inlets (injection station or dual station) and receive oil from multiple refineries. Once delivered or injected at any station along the pipeline route, the downstream flow rate may change. Multi-product pipelines transport multiple oil products with different oil properties. When different products are transported next to each other, oil mixing will inevitably occur between batches, and the tracking of oil mixing section and control of oil mixing volume is the key to optimal operation and management of multi-product pipelines, especially for pipelines laid in areas with complex terrain and large elevation differences.

In the case of a multi-product pipeline, an unreasonable batch sequence and injection volume during the planning process can lead to increased pollution from adjacent products being transported through the pipeline, which is a serious accident for actual production. At the same time, deviations in delivery times and quantities may lead to increased inventory costs or shortages of products in the warehouse. With the development of oil and gas pipelines, their topology is becoming more and more complex, the number of transported product batches is increasing, and the operating technology is becoming more flexible, which makes pipeline scheduling tends to be more difficult. The development of scheduling schemes through experience no longer meets the needs of today's pipeline system development. It has been a great concern to develop a reasonable pipeline scheduling plan that can meet the market demand and also ensure the safety and economy of pipeline operation.

1.2 Research Status

1.2.1 Optimization Model of Pipeline Network Scheduling

For oil and gas pipeline networks scheduling optimization problem, the key to establish the mathematical model is to determine the optimization objectives and constraints to be considered; moreover, both objectives and constraints will change with the actual needs of the project.

1.2.1.1 Gas Pipeline Network

For natural gas transmission pipelines, the scheduling models can be classified into the following types depending on the optimization objectives [6].

A. Lowest operational energy consumption. Compressor energy costs account for up to 50% of the operating and management costs of natural gas pipelines (networks), and there are still problems in the operation of pipelines (networks), such as unreasonable compressor start and stop, and reliance on manual experience for network flow distribution. In order to reduce the operating cost of natural gas pipeline network, an optimization model is often established with the objective function of minimizing the total energy consumption cost of the compressor for the whole line, and the pressure or flow control value of the compressor is determined according to the compressor power equation [7].
B. Maximum return on pipeline network operation. In the optimization of natural gas pipeline network with integrated production and sales, the optimization model is established with the objective function of maximizing the total revenue from the operation of the whole line, taking into account the purchase price of gas fields, the energy cost of pipeline transmission, and the price of gas sold by customers. Usually, the larger the transmission volume, the larger the revenue from gas sales, but also the larger the transmission energy consumption, so operators need to find the optimal transmission plan to achieve the maximum economic return. However, for independent pipeline companies, the midstream transportation of natural gas pipelines is separated from the upstream and downstream production and sales, so the revenue model for pipeline network operation does not need to consider the field purchase price and the customer sales price. The network operator needs to balance the revenue from gas transmission (instead of revenue from gas sales) and energy consumption to find the optimal gas transmission plan.
C. Maximum pipeline capacity. In order to fully utilize the capacity of the existing gas pipeline (network), an optimization model is often developed with the maximum total gas transmission capacity of the gas pipeline network as the objective function. The model needs to consider the topology of the pipeline network; pressure and flow constraints of the pipe section; pressure and flow

constraints of the injection and delivery nodes; compressor operation and other constraints to determine the optimal injection (delivery) volume of each injection and delivery node.

In the process of optimizing the natural gas pipeline scheduling problem, to determine the scheduling plan, it is necessary to consider several constraints such as the amount of dispatchable resources, demand of downstream users, and technology of natural gas pipeline network. The amount of natural gas resources entering the pipeline network in a limited period of time is constrained by production or processing capacity of gas wells, gathering stations, and purification plants due to the characteristics of natural gas extraction. At the same time, the pressure in the pipeline (or the total standard volume of natural gas) is only allowed to vary within a certain range due to the limitations of the process technology; in addition, the downstream users of natural gas need to meet certain requirements for the volume and pressure of gas delivered by the pipeline network due to their own needs, so the pressure and volume of gas delivered by the pipeline network cannot be lower than a certain limit; in order to meet the strength requirements of the pipeline during the transmission process, the pressure of gas delivered by the pipeline cannot be higher than a certain value.

1.2.1.2 Oil Pipeline Network

The multi-product pipeline network scheduling model can be classified into the following types according to the different optimization objectives [8–10].

A. Lowest operational energy consumption. In the process of oil batch transportation, both pumping station characteristics and pipeline characteristics will change significantly, thus causing changes in the operating point of the pumping station-pipeline system. At this point, the goal of the scheduler is to provide the optimal operating scheme for all pump stations along the entire line during the entire scheduling cycle on the basis of meeting the delivery volume. The entire program includes the opening and closing times of the equipment at each station along the pipeline, and specific speed values are provided for the adjustable speed pumps to ensure that the operating program meets the hydraulic and flow constraints and that the energy consumption of the pipeline operation is minimized.
B. Maximize return. According to the different sales prices of oil products and customer demand, establish the target revenue maximization objective function, calculate the difference between the product supply and energy consumption cost of the pipeline network. In order to obtain the maximum economic benefit, the pipeline network system needs to consider the cost of pump power consumption, as well as a comprehensive consideration of factors such as oil sources, coordination between delivery station depot inventories and the connection of each pipeline delivery batch.
C. Minimal download deviation. Since the oil demand is proposed by each delivery station individually, the actual situation of pipeline operation is not considered

when formulating, so the reasonableness of the plan cannot be guaranteed. If the oil demand of each station is set as an equation constraint, it may lead to the contradiction between model constraints. Part of the study takes the minimum deviation of each station from the actual delivery quantity of each oil product plan as the objective function, linearizes the objective function by introducing artificial relaxation variables and artificial tightening variables, and establishes the corresponding linear programming (LP) model.

D. Minimal oil mixing loss. In the process of multi-product pipeline sequential delivery, adjacent batches will inevitably form mixed oil segments under the effect of convection and turbulent diffusion. The presence of blending sections leads to a decrease in the amount of qualified product, which needs to be treated by additional processes, such as blending or distillation. Therefore, tracking of the blending section and control of the blending quantity is an important part of pipeline transportation that cannot be ignored.

E. Minimal deviation of delivery time window. For commercial multi-product pipeline serving the downstream market, the pipeline scheduling optimization model with the objective of minimum demand time window deviation can obtain a scheduling plan that better matches the demand time window proposed by the user and is more in line with the actual situation when constraints such as flow constraints, batch constraints, and demand time window constraints are considered.

For crude oil pipelines, the common optimization objectives are more similar to those for multi-product pipelines, including minimizing the number of switching between high and low freezing point crude oil transfers in the same pipeline, minimizing operating costs, maximizing return, and minimizing the delivery time deviation. It is worth noting that crude oil pipeline costs are different from those of multi-product pipelines, such as the fees paid by tankers to ports according to the length of their berthing time, the switching costs incurred when unloading crude oil to different port storage tanks, and the switching costs incurred when distillation towers need to switch between multiple supply tanks to feed them during refining. At the same time, there is a great difference in the flow pattern of crude oil transported in the pipe with different properties, for example, crude oil with high freezing point needs to be transported with heat, while crude oil with low freezing point can be transported at room temperature.

During establishing the scheduling model of the liquid pipeline network, it is also necessary to consider the constraints caused by the limitations of the pipeline system equipment and the operating process requirements. The first category is the flow constraints including injection flow, delivery flow, and pipe section flow, where injection flow and pipe section flow are mainly determined by the characteristics of centrifugal pumps at stations and the lower limit of mixing flow, and download flow is determined by the operating range of flow meters at delivery stations. The second category is pressure constraint, which mainly refers to the pressure limit of the whole line and key nodes (pumping stations, high points, etc.), and its interaction with flow control. The third category is the batch shipment constraint, in a crude oil or multi-product pipeline scheduling model with known time node sequencing, where the forward length of a batch head in a time window depends on the total number of downloads from all stations after it [11]. Usually, an initial scheduling plan can be developed based on the above three types of constraints. However, as the scheduling concept continues to develop in the direction of meticulous, digital, and intelligent, the plans solved by the above models often have problems such as large flow fluctuations, frequent pump starts and stops, and inconvenient management, thus not serving the actual needs of the site well. Therefore, for different pipeline models, different additional constraints need to be considered, such as the problem of oil mixing between adjacent batches. Certainly, the process of crude oil pipeline scheduling model establishment also requires additional consideration of port berth constraints, depot operation constraints, tank capacity constraints, etc. [12, 13].

In addition, it is also necessary to consider which time and volume expressions are used to describe the transportation process of liquid batches in the pipeline. According to the different time and volume expressions, they can be divided into the discrete time and volume-based model (RR model for short) proposed by Rejowski, and the continuous time and volume model (CC model for short) first proposed by Cafaro. In the RR model, the scheduling period is discretized into a number of equally spaced time steps according to the specified time step, and the start and end time of each time step are known. And each pipe section along the pipeline is discretized into a number of micro-element segments according to the specified volume step, and each micro-element segment is required to contain only one oil product at each time. The CC model selects the starting and ending moments of each batch at the pipeline start point as the time nodes and discretizes the scheduling period into several time steps. Since the start and end moments of each batch at the pipeline start input are decision variables, the start and end moments of each time step are also decision variables.

1.2.2 Solution Algorithms

Regarding the optimization problem of oil and gas pipeline scheduling, there are five main solution strategies as follows.

A. Establish a complete optimization model and solve the model directly by using classical optimization methods (such as simplex method, cut-plane method, branch-and-bound method, Newton iteration method, dynamic programming) or commercial optimization solvers (such as CPLEX, Gurobi, etc.) [14]. Among them, dynamic programming is a common method to solve the scheduling optimization problem of branching natural gas pipeline networks [15]. The advantage of classical optimization methods is that the global optimality of the solution is guaranteed, but this exact solution method with global search can also make the time and space complexity of the solution increase significantly. For example, the scheduling model can suffer from dimensional catastrophe when there are large-scale integer variables in the model, and then it is impossible to find a feasible solution in polynomial time. Therefore, for multivariate scheduling, a single mathematical planning model often cannot cover all the factors and has problems such as large solution space and computational difficulties.

B. Establish a complete optimization model and solve the model in two stages: The first stage is model pre-processing, specifically by using heuristic methods such as genetic algorithm, ant colony algorithm, and simulated annealing algorithm to determine the values of some variables in the model, such as determining the order of oil inject at the starting point of the pipeline through pre-processing; the second stage is solving the pre-processed model using classical optimization methods. In the field of oil and gas pipeline network scheduling, heuristic algorithms are proposed to greatly reduce the time required to solve the scheduling plan. In the solution process, these two stages can be completed either sequentially and at once or by coupling [16]. The research results show that the hybrid heuristic method is effective for solving the mathematical basis problems such as batch sequencing nonlinear problems, hydraulic scheduling nonlinear coupling problems, and the efficiency of large-scale model solving. However, the use of this method inevitably leads to problems such as poor optimality of solution results and poor generalizability of the method. Therefore, the method is less practical for scheduling problems that are more sensitive to the objective function and for scheduling problems with a special pipeline structure.

C. Instead of directly establishing a strict mathematical model for the problem under study, the original problem is solved directly using heuristic methods or intelligent algorithms oriented to the optimization objectives and constraints [14, 17].

D. The decomposition strategy is used to decompose the original problem into several subproblems, establish a mathematical model for each subproblem, and use classical optimization methods or commercial optimization solvers to solve each submodel [18, 19].

E. Decomposition strategy is used to decompose the original problem into several subproblems, model some of them, and solve them using classical optimization methods or commercial optimization solvers, while the other subproblems are solved directly using heuristic methods or intelligent algorithms [20].

The scheduling results derived from the initial research methods focusing on mathematical programming and heuristic algorithms for scheduling problems were usually not satisfactory enough to solve practical scheduling problems. With the establishment and development of various related disciplines and optimization techniques, many new means and methods have emerged in the scheduling field, such as various scheduling methods based on deep learning and self-learning. The emergence of these methods has enriched the means of solving scheduling problems and led to the development of research on scheduling problems in a diversified direction. At present, a variety of methods have been applied to solving scheduling problems of large-scale and complex systems.

1.3 Conclusion

1.3.1 General Situation

In recent years, although many countries and regions have started to develop renewable energy sources, this has led to the replacement of some fossil energy sources by renewable energy sources and a gradual increase in the proportion of the world's energy consumption. However, according to the International Energy Agency (IEA), oil and natural gas will still dominate the future energy consumption [21]. It can be inferred from this that the rational arrangement of oil and gas resources remains the focus of pipeline construction.

Currently, the available results lack a comprehensive review from the perspective of the development of pipeline scheduling problems. In this section, a bibliometric approach is used for the case of multi-product pipelines, and the research shortcomings and trends are analyzed to provide an idea of some directions and inspirations for future research on pipeline scheduling. We selected Web of Science (WOS) and China National Knowledge Infrastructure (CNKI) as data sources and counted the number of articles related to the scheduling of multi-product pipelines in the above two databases, and the results are shown in Fig. 1.3. From 2000 to March 2022, 250 articles have been published cumulatively, and the research fervor of them still remains at a high level. Furthermore, the statistics for the types of articles in the area of multi-product pipeline scheduling and the journals in which they were published are shown in Fig. 1.4. For each area in the figure, the number above represents the number of published articles and the number below represents the percentage. As can be seen in Fig. 1.4, the largest number of research articles, 72%, and the smallest number of review articles, 3%, are published. Also, we further counted the journals that published research articles and review articles. It can be seen that articles on multi-product pipeline scheduling are mainly published in journals of petroleum, chemical engineering, and operations research, among which Computer and Chemical Engineering, Oil and Gas Storage and Transportation, and Industrial & Engineering Chemistry Research have the largest number of articles in this field, accounting for 17%, 14%, and 12%, respectively [22].

Next, we analyzed the publishing institutions of the collected articles, and it can be seen that the main countries studying this issue include Argentina, Brazil, China, Finland, Iran, Portugal, Spain, and the USA. Most of the researchers focus on the short-term scheduling of small-scale pipeline systems, while some extend their studies to the long-term scheduling of small-scale pipeline systems or the short-term scheduling of large-scale pipeline systems.

The literature collected above was further analyzed to summarize the research framework developed in the past, as shown in Fig. 1.5 [22]. First, the scheduling model of the pipeline needs to be determined. The scheduling models of pipelines can be divided into deterministic scheduling and scheduling under uncertain environments [23, 24]. If different models are chosen, the model structure and decision variables involved in the scheduling model are different (see Sect. 1.1.2). After the

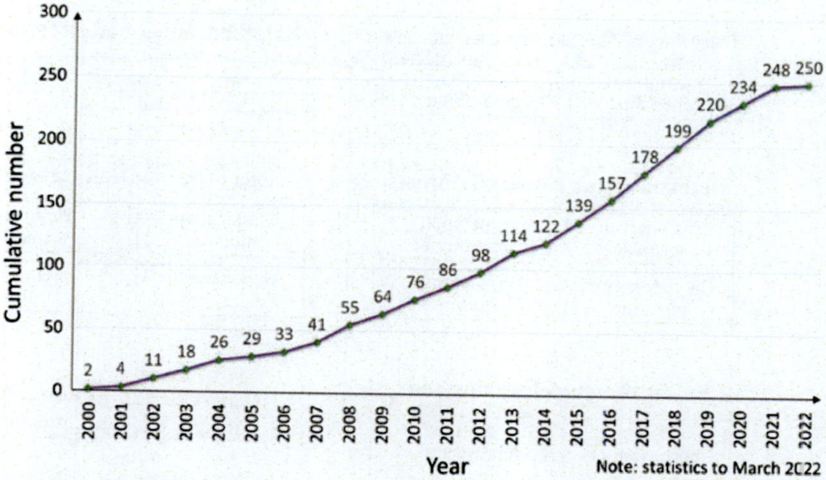

Fig. 1.3 Cumulative publication number of articles related to multi-product pipeline scheduling

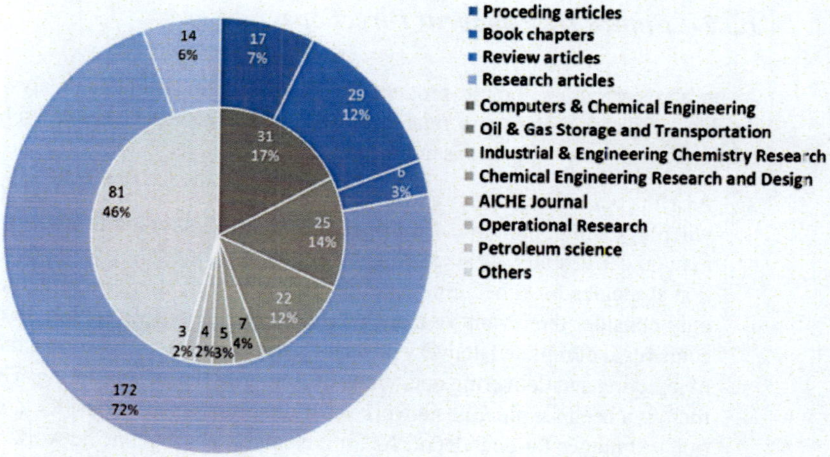

Fig. 1.4 Types of articles and journals related to multi-product pipeline scheduling

model is selected, the object of study, such as the topology of the pipeline, operational constraints, and scheduling accuracy, needs to be determined, and then a suitable modeling method should be selected to build the scheduling model (see Sect. 1.2.1) [19, 25]. Finally, a suitable solution strategy should be selected according to the model structure (see Sect. 1.2.2).

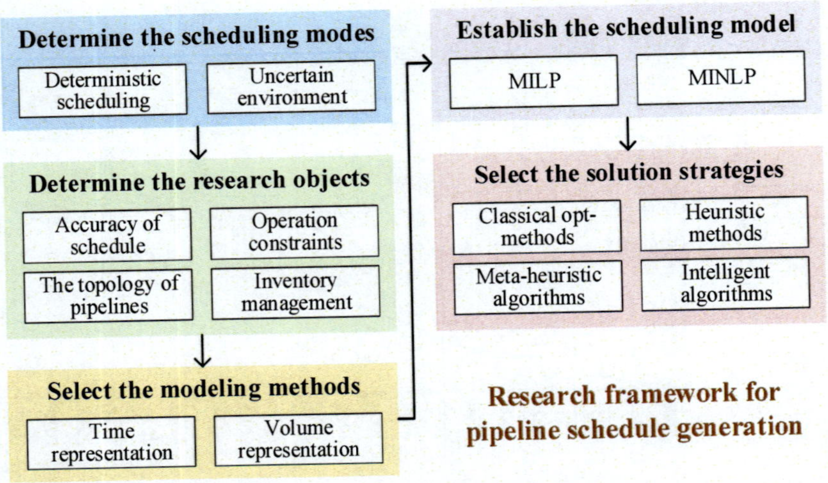

Fig. 1.5 Research framework of multi-product pipeline scheduling

1.3.2 Future Development Directions

As the above example of multi-product pipeline scheduling literature analysis, based on the comparison of papers related to oil and gas pipeline network scheduling optimization, some suggestions are made for future research.

A. As a link between upstream and downstream liquid pipelines, dual depot (station) will play an important role in the operation of crude oil or multi-product pipeline networks. Currently, most existing studies focus on modeling methods and solution strategies for scheduling pipeline networks with complex structures. They only consider the inventory management of dual depot, or setting up different centrifugal pumps and delivery flows for specific products, but ignore the impact of injecting and delivering operations of storage tanks on scheduling. Therefore, there is a need for pipeline network scheduling studies that consider tank operation techniques for dual depot. As for the natural gas pipeline network, it is also necessary to combine the injection and extraction of gas storage and utilization of line-pack gas with the scheduling of the pipeline network to play a better peak regulation effect.
B. The improvement of interconnection between pipelines has intensified the difficulty of pipeline network scheduling. At present, most effective scheduling methods are applicable to small-scale pipeline networks and cannot meet the timeliness requirements of large-scale pipeline network scheduling. Therefore, an efficient solving method based on machine learning can be developed in conjunction with the historical scheduling of pipelines, so as to realize the rapid preparation of large-scale pipeline network scheduling.

C. At present, the existing research mainly focuses on static scheduling, while the research on dynamic scheduling is still in the preliminary stage and limited to the rolling generation of linear pipeline summary scheduling. In the future, research on dynamic scheduling and online scheduling can be further developed. Based on the historical data of inventory in the depot, the product demand in the depot can be accurately predicted. At the same time, combined with real-time data transmitted by supervisory control and data acquisition (SCADA) systems, dynamic and online scheduling models can be built and combined with trend forecasts of product supply and demand. With this model, inventory risk can be dynamically assessed on a rolling cycle and optimal detailed schedules can be generated. The method can provide dynamic forecasting, risk assessment, result feedback, and global adjustment for the pipeline network and can be a decision-making tool for intelligent scheduling of oil and gas pipeline networks.

D. Oil and gas pipeline systems are highly uncertain, which will make deterministic scheduling no longer feasible. Considering robustness in the scheduling model can reduce the risk caused by uncertainties and the number of scheduling adjustments, so that scheduling can still meet market requirements within a certain demand range. However, most of the existing research focuses on oil and gas supply chain optimization under uncertainties, while the research on robustness optimization of pipeline scheduling is focused on pipeline systems with single sources and distribution centers. There is a need to extend the research to pipeline systems with complex structures by incorporating uncertainty optimization theory.

References

1. Q. Liao, Y. Liang, N. Xu, H. Zhang, J. Wang, and X. Zhou, "An MILP approach for detailed scheduling of multi-product pipeline in pressure control mode," *Chemical Engineering Research and Design,* vol. 136, pp. 620–637, 2018/08/01/ 2018, doi: https://doi.org/10.1016/j.cherd.2018.06.016.
2. X. Zhou *et al.*, "Future scenario of China's downstream oil supply chain: Low carbon-oriented optimization for the design of planned multi-product pipelines," *Journal of Cleaner Production,* vol. 244, p. 118866, 2020/01/20/ 2020, doi: https://doi.org/10.1016/j.jclepro.2019.118866.
3. X. Zhou *et al.*, "A hybrid time MILP model for the pump scheduling of multi-product pipelines based on the rigorous description of the pipeline hydraulic loss changes," *Computers & Chemical Engineering,* vol. 121, pp. 174–199, 2019/02/02/ 2019, doi: https://doi.org/10.1016/j.compchemeng.2018.10.001.
4. L. Fan *et al.*, "A deep reinforcement learning-based method for predictive management of demand response in natural gas pipeline networks," *Journal of Cleaner Production,* vol. 335, p. 130274, 2022/02/10/ 2022, doi: https://doi.org/10.1016/j.jclepro.2021.130274.
5. L. Yu, S. Wang, and Q. Xu, "Optimal scheduling for simultaneous refinery manufacturing and multi oil-product pipeline distribution," *Computers & Chemical Engineering,* vol. 157, p. 107613, 2022/01/01/ 2022, doi: https://doi.org/10.1016/j.compchemeng.2021.107613.

6. H. Su *et al.*, "A method for the multi-objective optimization of the operation of natural gas pipeline networks considering supply reliability and operation efficiency," *Computers & Chemical Engineering,* vol. 131, p. 106584, 2019/12/05/ 2019, doi: https://doi.org/10.1016/j.compchemeng.2019.106584.
7. M.-w. Fan, C.-c. Ao, and X.-r. Wang, "Comprehensive method of natural gas pipeline efficiency evaluation based on energy and big data analysis," *Energy,* vol. 188, p. 116069, 2019/12/01/ 2019, doi: https://doi.org/10.1016/j.energy.2019.116069.
8. Y. Liang, M. Li, and J. Li, "Hydraulic model optimization of a multi-product pipeline," *Petroleum Science,* vol. 9, no. 4, pp. 521–526, 2012/12/01 2012, doi: https://doi.org/10.1007/s12182-012-0237-2.
9. Z. Haoran, L. Yongtu, L. Qi, S. Yun, and Y. Xiaohan, "A self-learning approach for optimal detailed scheduling of multi-product pipeline," *Journal of Computational and Applied Mathematics,* vol. 327, pp. 41–63, 2018/01/01/ 2018, doi: https://doi.org/10.1016/j.cam.2017.05.040.
10. Q. Liao, H. Zhang, N. Xu, Y. Liang, and J. Wang, "A MILP model based on flowrate database for detailed scheduling of a multi-product pipeline with multiple pump stations," *Computers & Chemical Engineering,* vol. 117, pp. 63–81, 2018/09/02/ 2018, doi: https://doi.org/10.1016/j.compchemeng.2018.05.002.
11. Q. Liao, P. M. Castro, Y. Liang, and H. Zhang, "New batch-centric model for detailed scheduling and inventory management of mesh pipeline networks," *Computers & Chemical Engineering,* vol. 130, p. 106568, 2019/11/02/ 2019, doi: https://doi.org/10.1016/j.compchemeng.2019.106568.
12. V. G. Cafaro, P. C. Pautasso, J. Cerdá, and D. C. Cafaro, "Efficient planning of crude oil supplies through long-distance pipelines," *Computers & Chemical Engineering,* vol. 122, pp. 203–217, 2019/03/04/ 2019, doi: https://doi.org/10.1016/j.compchemeng.2018.06.028.
13. Y. Hou, Y. Zhang, N. Wu, and Q. Zhu, "Constrained multi-objective optimization of short-term crude oil scheduling with dual pipelines and charging tank maintenance requirement," *Information Sciences,* vol. 588, pp. 381–404, 2022/04/01/ 2022, doi: https://doi.org/10.1016/j.ins.2021.12.067.
14. V. G. Cafaro, D. C. Cafaro, C. A. Méndez, and J. Cerdá, "Detailed Scheduling of Single-Source Pipelines with Simultaneous Deliveries to Multiple Offtake Stations," *Industrial & Engineering Chemistry Research,* vol. 51, no. 17, pp. 6145–6165, 2012/05/02 2012, doi: https://doi.org/10.1021/ie202520a.
15. Y.-Z. Meng, R.-R. Chen, and T.-H. Deng, "Two-stage robust optimization of power cost minimization problem in gunbarrel natural gas networks by approximate dynamic programming," *Petroleum Science,* 2022/01/13/ 2022, doi: https://doi.org/10.1016/j.petsci.2021.09.048.
16. S. Relvas, A. P. F. D. Barbosa-Póvoa, and H. A. Matos, "Heuristic batch sequencing on a multiproduct oil distribution system," *Computers & Chemical Engineering,* vol. 33, no. 3, pp. 712–730, 2009/03/20/ 2009, doi: https://doi.org/10.1016/j.compchemeng.2008.10.012.
17. S. A. MirHassani and N. BeheshtiAsl, "A heuristic batch sequencing for multiproduct pipelines," *Computers & Chemical Engineering,* vol. 56, pp. 58–67, 2013/09/13/ 2013, doi: https://doi.org/10.1016/j.compchemeng.2013.05.007.
18. L. Magatão, L. V. R. Arruda, and F. Neves, "A mixed integer programming approach for scheduling commodities in a pipeline," *Computers & Chemical Engineering,* vol. 28, no. 1, pp. 171–185, 2004/01/15/ 2004, doi: https://doi.org/10.1016/S0098-1354(03)00165-0.
19. V. G. Cafaro, D. C. Cafaro, C. A. Méndez, and J. Cerdá, "MINLP model for the detailed scheduling of refined products pipelines with flow rate dependent pumping costs," *Computers & Chemical Engineering,* vol. 72, pp. 210–221, 2015/01/02/ 2015, doi: https://doi.org/10.1016/j.compchemeng.2014.05.012.
20. H. Zhang, Y. Liang, Q. Liao, M. Wu, and X. Yan, "A hybrid computational approach for detailed scheduling of products in a pipeline with multiple pump stations," *Energy,* vol. 119, pp. 612–628, 2017/01/15/ 2017, doi: https://doi.org/10.1016/j.energy.2016.11.027.
21. T. Kober, H. W. Schiffer, M. Densing, and E. Panos, "Global energy perspectives to 2060 – WEC's World Energy Scenarios 2019," *Energy Strategy Reviews,* vol. 31, p. 100523, 2020/09/01/ 2020, doi: https://doi.org/10.1016/j.esr.2020.100523.

22. Z. Li, Y. Liang, Q. Liao, B. Zhang, and H. Zhang, "A review of multiproduct pipeline scheduling: from bibliometric analysis to research framework and future research directions," *Journal of Pipeline Science and Engineering,* vol. 1, no. 4, pp. 395–406, 2021/12/01/ 2021, doi: https://doi.org/10.1016/j.jpse.2021.08.001.
23. X. Dai *et al.*, "A data-driven approach for crude oil scheduling optimization under product yield uncertainty," *Chemical Engineering Science,* vol. 246, p. 116971, 2021/12/31/ 2021, doi: https://doi.org/10.1016/j.ces.2021.116971.
24. C. Lima, S. Relvas, and A. Barbosa-Póvoa, "Stochastic programming approach for the optimal tactical planning of the downstream oil supply chain," *Computers & Chemical Engineering,* vol. 108, pp. 314–336, 2018/01/04/ 2018, doi: https://doi.org/10.1016/j.compchemeng.2017.09.012.
25. Q. Liao, P. M. Castro, Y. Liang, and H. Zhang, "Computationally Efficient MILP Model for Scheduling a Branched Multiproduct Pipeline System," *Industrial & Engineering Chemistry Research,* vol. 58, no. 13, pp. 5236–5251, 2019/04/03 2019, doi: https://doi.org/10.1021/acs.iecr.8b06490.

Chapter 2
Advanced Modeling and Algorithm for Pipeline Scheduling

Ning Xu and Bo Zhang

Abstract In practice, the scheduling of gas and oil pipeline networks is a continuous and dynamic process with a large number of uncertainties. Therefore, shorter calculation time and solution quality are always important indicators to measure the practicability of solution methods, especially in the real-time and active pipeline scheduling. The emergence of digital and intelligent technologies makes the data sensing and decision-making possible to be real-time and continuous. This chapter summarizes existing research from the perspective of modeling methods and solution algorithms, so as to provide the research basis and research direction for intelligent scheduling of pipeline networks. The modeling methods are mainly divided into mathematical programming, generalized disjunctive programming (GDP), and resource task network (RTN). The solution algorithms include mathematical programming, heuristic algorithm, metaheuristic algorithm, dynamic programming algorithm, and data-driven algorithm.

2.1 Modeling Methods

For pipeline scheduling, the most common modeling method at present is to build a mathematical programming model. By determining the objective function and constraints, the obtained results can meet the actual needs, and the optimal solution can be obtained through the theory of operations research. However, this method has some shortcomings, such as the difficulty of solving when the scale is large and the difficulty of describing the problem in some cases. Therefore, some scholars have proposed modeling methods of GDP and RTN.

N. Xu (✉)
China University of Petroleum-Beijing at Karamay, Xinjiang 834000, PR China
e-mail: xn11235@cup.edu.cn

N. Xu · B. Zhang
National Engineering Laboratory for Pipeline Safety/MOE Key Laboratory of Petroleum Engineering/Beijing Key Laboratory of Urban Oil and Gas Distribution Technology, China University of Petroleum-Beijing, Fuxue Road No.18, Changping 102249, Beijing, PR China
e-mail: zb131211@163.com

H. Su et al. (eds.), *Advanced Intelligent Pipeline Management Technology*,
https://doi.org/10.1007/978-981-19-9899-7_2

2.1.1 Mathematical Programming

2.1.1.1 Natural Gas

The optimization of natural gas pipeline network scheduling plan is divided into two categories of steady state and non-steady state. Steady-state scheduling optimization refers to the problem under the condition that the operating conditions of the whole pipeline do not change with time. But fluctuations in gas consumption, compressor start and stop, valve adjustment, leakage or rupture of pipelines, etc., will cause the pipeline to operate in an unsteady state. In this case, the operating parameters are highly coupled with time and space. The non-steady-state operation optimization of natural gas pipeline network system is to optimize flow rate, pressure, and compressor operating conditions of each node in the system in real time when the system is in a non-steady state, so that the pipeline network system is always safe and operate under high-efficiency conditions.

(1) Steady-state scheduling

The current difficulties in the optimization of steady-state scheduling plan lie in the complicated topology of the annular gas transmission network, the uncertainty of gas flow direction, and the changing distribution of transmission volumes, etc. As a result, non-sequential dynamic programming (NDP), generalized parsimonious gradient method, heuristic algorithms, stochastic optimization algorithms, artificial intelligence algorithms, and other methods are frequently used to separate and simplify pipeline network systems.

The objective of gas network schedule optimization is to discover the most cost-effective transmission solution by reducing gas transmission energy consumption while guaranteeing that gas transmission jobs are completed. Given the pipeline capacity and other uncontrollable elements, the ideal operation mode, equipment combination, and operating parameters are required to minimize energy consumption per unit of time under steady-state conditions, subject to pipeline operation limitations. The constraints of gas transmission pipelines steady-state operation include inlet or delivery gas volume constraints, inlet or delivery gas pressure constraints, pipeline pressure drop equation constraints, pipeline network node flow balance constraints, compressor power constraints, compressor equation constraints, valve equation constraints, etc.

The volume of gas delivered into the gas pipeline is necessarily influenced by production fluctuations at different times at gas production sites, as well as pressure constraints and requirements for gas purchased externally by gas operators. Furthermore, due to their own needs, individual customers have specific requirements for the volume and pressure of gas supplied to the pipeline network, i.e., the pressure at the network's distribution points cannot go below a set pressure value. Excessive pressure may cause damage to the pipeline when natural gas enters the pipeline. The total amount of gas entering the network should equal the total amount of gas exiting the network to provide a balanced supply of gas across the network. The flow rate of gas pipeline can be calculated by Eq. 2.1, in which q_{ij} represents the mass volume rate value from node i to node i, C_{ij} represents the coefficient of frictional resistance from node i to node j, and e^s represents the elevation coefficient form node i to node j.

$$q_{ij} = C_{ij}\sqrt{p_i^2 - e^s p_j^2} \tag{2.1}$$

When the flow rate of a centrifugal compressor is decreased to a specific level, the consequent rotation and detachment process might result in fluid backflow. The gas pressure in the compressor outlet line is reduced once backflow occurs, and the compressor outlet pressure gradually rises with the backflow. The fluid will flow backward when the pressure reaches a specific level. Centrifugal compressor surge is a process that repeats itself again and over. If the centrifugal compressor's flow rate reaches a particular point, the gas flow rate in the compressor may exceed the speed of sound, and the flow rate will not increase again—phenomenon known as centrifugal compressor stall. The stagnation flow rate is the highest flow rate possible. If the gas flow of a centrifugal compressor exceeds its hysteresis flow, the hysteresis phenomenon may occur, resulting in energy waste. The compressor flow rate must therefore be lower than the stalling flow rate.

In engineering, compressor characteristic curves are often used to characterize compressor operation. A quadratic polynomial is used to simulate each curve on a centrifugal compressor characteristic diagram, as shown by Eq. 2.2. Equation 2.3 shows that a double quadratic polynomial can also be used to simulate the centrifugal compressor characteristic curve. The coefficients $a \in \mathbb{R}^3$ and $A \in \mathbb{R}^{3\times3}$ are based on technical measurement data for a given compressor, fitted by the least squares method. The contour of the compressor speed $n \in \left[n^{\min}, n^{\max}\right]$ is given by the implicit Eq. 2.4, in which n represents the speed of the compressor, and A^{speed} represents the experimentally obtained fitting constant.

$$\psi(x, A) = a_0 + a_1x + a_2x^2 \tag{2.2}$$

$$\chi(x, y, A) = \begin{pmatrix} 1 \\ x \\ x^2 \end{pmatrix}^T \begin{pmatrix} a_{0,0} & a_{0,1} & a_{0,2} \\ a_{1,0} & a_{1,1} & a_{1,2} \\ a_{2,0} & a_{2,1} & a_{2,2} \end{pmatrix} \begin{pmatrix} 1 \\ y \\ y^2 \end{pmatrix} \tag{2.3}$$

$$H_{\text{ad}} = \chi\left(Q, n, A^{\text{speed}}\right) \tag{2.4}$$

(2) Non-steady-state scheduling

The variables of non-steady-state scheduling optimization for natural gas pipeline system including the pressure of compressors at each time and the optimization objective are integration of energy consumption at each time. The discretization of the time factor in the model increases the number of model variables significantly, making the model more difficult to solve due to its nonlinear, combinatorial, and stochastic nature.

The restrictions in the mathematical model for optimizing natural gas pipeline system dynamic operation differ slightly from those in the mathematical model for optimizing natural gas pipeline network steady-state operation. Some of the limitations consider the impact of time on the constraints. The mathematical equations for the dynamic operation of natural gas in a pipeline differ significantly from the mathematical description of its steady-state flow. If no changes in the temperature of the natural gas along the pipeline are considered, the dynamic flow of natural gas in the pipeline can be described by the continuity equation and the equation of motion; if changes in temperature along the pipeline are considered, the dynamic flow of natural gas in the pipeline should be described by the continuity equation, the equation of motion, and the energy equation together. Equations 2.5 and 2.6 are the continuity equation and momentum equation, respectively, in which A represents the cross-sectional area of the pipe, ρ represents the gas density, τ represents time, w represents the gas flow rate, p represents the gas pressure, g represents the acceleration of gravity, x represents for location, and λ represents the coefficient of friction.

$$A\frac{\partial \rho}{\partial \tau} + \frac{\partial}{\partial x}(\rho w A) = 0 \tag{2.5}$$

$$\frac{\partial(\rho w)}{\partial \tau} + \frac{\partial p}{\partial x} + \frac{\partial(\rho w^2)}{\partial x} = -g\rho\frac{\mathrm{ds}}{\mathrm{dx}} - \frac{\lambda}{d}\frac{w^2}{2}\rho \tag{2.6}$$

Natural gas's uneven flow in a pipeline is a slow transient process, unlike that of a liquid, due to its compressible nature. As a result, the two equations above can be differentiated using a central implicit differencing method with less time step requirements and without taking temperature swings into account. Since the time step can be as long or as short as desired during the simulation, it increases the simulation's flexibility while simultaneously reducing the simulation's processing time. The central implicit difference continuity equation, as well as the momentum equation, is shown by Eqs. 2.7 and 2.8.

$$\rho_i^{k+1} - \rho_i^k + \rho_{i+1}^{k+1} - \rho_{i+1}^k + \gamma(m_{i+1}^k - m_i^k + m_{i+1}^{k+1} - m_i^{k+1}) = 0 \tag{2.7}$$

$$\begin{aligned} m_i^{k+1} &- m_i^k + m_{i+1}^{k+1} - m_{i+1}^k + \gamma[p_{i+1}^k + (m_{i+1}^k)^2/\rho_{i+1}^k - p_i^k - (m_i^k)^2/\rho_i^k \\ &+ p_{i+1}^{k+1} + (m_{i+1}^{k+1})^2/\rho_{i+1}^{k+1} - p_i^{k+1} - (m_i^{k+1})^2/\rho_i^{k+1}] \\ &+ \frac{1}{2}\Delta\tau g\frac{ds}{dx}(\rho_i^{k+1} + \rho_i^k + \rho_{i+1}^{k+1} + \rho_{i+1}^k) \\ &+ \frac{1}{4}(\Delta\tau\lambda/d)[(m_i^{k+1})^2/\rho_i^{k+1} + (m_i^k)^2/\rho_i^k + (m_{i+1}^{k+1})^2/\rho_{i+1}^{k+1} + (m_{i+1}^k)^2/\rho_{i+1}^k] = 0 \end{aligned} \tag{2.8}$$

2.1.1.2 Refined Oil

Refined oil pipeline scheduling is the process of transporting refined oil from supply to demand, which can be divided into static scheduling and dynamic scheduling. The static scheduling is based on the injection plan of injection stations within a certain time interval. According to the injection time of each batch and pipeline flow rate, the time when each batch arrives at each delivery station can be obtained, so as to determine the delivery flow rate and time of each station along the pipeline for each batch. The dynamic scheduling is to determine the injection flow rate and time of each batch at injection stations based on the inventory of injection stations, and the demand volume and time of delivery stations along the pipeline.

(1) Static scheduling

For static scheduling, injection stations must inject refined oil according to predetermined time, and the pipeline must operate in specific flow rate to ensure that refined oil can be delivered on time. In this way, the delivery time can only be determined by the carriers, and shippers can't determine the arrival time of refined oil. In the process of batch transportation, unforeseeable operating conditions will also affect the delivery time and delivery volume, such as equipment failure maintenance, the storage capacity of the delivery stations, the supply, etc. Especially, when customers only purchase refined oil that is urgently needed, it will cause customers to encounter production and operational problems because delivery time is often inflexible. This method is advantageous when shippers only need to track the individual batches.

Aiming at the problem about optimizing batch schedules of a refined oil pipeline, the established models can be divided into two types according to the scheduling model time representation expression, namely discrete and continuous time representation. For a detailed batch scheduling plan, there are many key time nodes to mark the operations to be performed. For example, a station starts or stops to delivery refined oil, or a pipeline section stops oil transportation, etc. Therefore, key time nodes are critical for field operators. We can divide the batch scheduling plan into multiple parts through key time nodes. Each part corresponds to a time window. The field operators can clearly understand the running status of the pipeline in each time window, including the delivery flow rate of stations, the transporting status of pipeline segments, etc.

(a) Discrete time representation

For discrete time representation, the scheduling period is divided into several time windows of equal or specified length. Magatão et al. [1] made full use of the advantages of discrete time representation and used this representation to analyze the pipeline network scheduling problem between ports and refineries. Herrán et al. [2] successfully solved the problem about tracking of batch interface in the case of multiple injection stations and distribution stations. However, for discrete time representation, the model solution results will not be refined enough and lack practicability if the time window is too long. Conversely, too short a time window will result in a model that is too large to obtain a feasible solution in an acceptable time.

Therefore, continuous time representation has been extensively studied to solve the above-mentioned problems.

(b) Continuous time representation

The continuous time representation uses key events as the standard to select time nodes, such as the time of starting or stopping to delivery, or the shutdown of a refined oil pipeline, etc. So that the model scale can be reduced as much as possible and the solution efficiency can be improved. Cafaro and Cerdá [3] established a mixed integer linear programming model based on continuous time representation for a refined oil pipeline with an injection station and multiple distribution stations. The model takes the minimum sum of the pump energy cost, the treatment cost of the oil mixture, and the inventory cost of the refined oil depot as the objective function. On this basis, Cafaro and Cerdá [4] took a branch-shaped refined oil pipeline network as the research object and established a mixed integer linear programming model based on continuous time representation. The model strictly tracks batches, allowing products to be injected and distributed at the same time. However, this model belongs to the rough scheduling model, and the batch scheduling plan obtained by the model can only provide the approximate time range of the injection/distribution of products and cannot determine the detailed operation time in each station. Some scholars took the single-source multi-sink refined oil pipeline as the research object, carried out a series of studies on the careful scheduling of the refined oil pipeline, and obtained the injection/distribution operation and the specific batch injection/ distribution volume of each station in each time period. Haoran Zhang et al. [5, 6] studied the batch arrival time, the start time of batch download at each station, and the variable flow operation time of the first station as event points. The original model is converted into time nodes sequencing issue and a mixed-integer linear programming (MILP) model which is based on known time node sequence. To improve the solution efficiency, a self-learning approach is proposed based on fuzzy clustering analysis which can quickly find good time sequence. In addition, some scholars have done some research on rolling scheduling of refined oil pipelines.

(2) Dynamic scheduling

Dynamic scheduling is suitable for the situation in which shippers have special requirements on the arrival time of refined oil. The method can supply refined oil to the delivery stations within the specified time interval.

The dynamic scheduling and static scheduling of refined oil pipelines have the same advantages. For example, it is convenient to track individual batches through a pipeline system. However, dynamic scheduling has disadvantages. For example, the successful execution of the scheduling plan depends on whether the refined oil is injected within the specified time. If a certain batch cannot be injected on time, the injection and delivery time of subsequent batches will change, making it impossible to meet customers' demand. At the same time, the transportation time is also long, and the time for each batch to arrive at each delivery station is also easy to change. The formulation of the dynamic refined oil pipeline scheduling plan will become more and more difficult in the long-distance refined oil pipeline.

In addition, the emergence of cyber-physical systems (CPS) and big data theory provide new opportunities and challenges for distributing products actively [7]. Compared to traditional passive scheduling pattern, the CPS-based one can use big data technology to discover valuable information in a wealth of data from CPS and forecast downstream demand accurately. Then the operators can deliver the predicted products to the nearest delivery point in advance, thus reducing the transportation time and improving delivery efficiency and customer satisfaction. The key for CPS-based scheduling technology is to ensure the accuracy of the demand forecasting as well as the computational efficiency of the scheduling optimization method. Nowadays, there are some mature methods for demand forecasting, such as autoregressive integrated moving average model (ARIMA), artificial neural network (ANN), support vector machine (SVM), and the improved methods of the above. While the traditional optimization methods for pipeline scheduling reveal several defects in face of CPS. First, experience-based methods mostly apply to the issues with clear and fixed boundary conditions, so their optimality degrades obviously if given the continuous, dynamic, and complex boundary conditions from CPS. Second, the common solution used in the model-based methods is the global searching method, the solution efficiency of which drops sharply once the model scale reaches a certain size. Nowadays, CPS-based scheduling technology has been carried out for several years in vehicle scheduling, but relative research on pipeline scheduling is still rare. Unlike vehicle scheduling, operations of pipeline scheduling are rather complicated, which requires in-depth research.

2.1.2 Generalized Disjunctive Programming (GDP)

Most scheduling optimization problems can be solved through constructing mathematical programming models, but solution efficiency of this method will greatly reduce when facing large-scale problems because it contains large number of binary variables and big-M constraints, and they will greatly increase the solution domain [8]. In order to deal with this problem, a framework based on equations and symbolic logic developed [9] which can convert the mathematical programming model into GDP model which is constructed by Boolean variables. Through GDP model, the problem can be expressed clearly and the large solution domain can be decreased by adopting logical reasoning. Then, the solution efficiency can be greatly improved. Mostafaei and Castro [10] proposed a GDP model using continuous time representation for determining the detailed plan of a pipeline with multi injection stations and delivery stations. In this way, the obtained plan can be determined in short time and pipeline utilization has been greatly improved. The model was later extended into a product-centric model based on GDP [11]. The binary variables in the GDP model are only one quarter of that in RTN model to be introduced.

2.1.3 Resource Task Network (RTN)

RTN is a network which contains resource nodes and task nodes. In the network, resource nodes represent actual objects such as equipment for processing or storage, and task nodes represent technologies that can convert one resource into another. Usually, RTN model is relatively simple, and it can be used for general scheduling systems. In addition, when applying RTN model into pipeline scheduling problems, continuous time representation and discrete time representation are also adopted. For example, Castro [12] proposed a RTN model based on continuous time representation for a pipeline network. In the problem, refined oil, refineries, pipelines, and depots are taken as resource nodes, while injection and delivery are taken as task nodes. However, when establishing RTN models for large-scale problems with complicating operational constraints, the process will be extremely complex, resulting in poor practicability.

2.2 Solution Algorithms

For the models constructed above, many scholars have proposed corresponding solution algorithms. For example, for mathematical programming models, mature mathematical programming methods can be used to solve them directly. Heuristic algorithms or metaheuristic algorithms can also be used to obtain partial solutions of mathematical programming models in advance, thereby reducing the difficulty of directly solving models. In addition, with the rapid development of machine learning, big data is gradually introduced into solving pipeline scheduling problems to improve solution quality and solution efficiency.

2.2.1 Mathematical Programming

Mathematical programming is a method of operational research that relies on the traditional search for global optimal solution of the planning algorithm to solve the pipeline scheduling model, in view of the large-scale scheduling optimization problems, commonly solved by the professional solving software or solver. The suitable solving software or solver can reduce the computation time and improve solution efficiency. At present, the popular solving software (solver) includes Lingo software developed by Lindo Company, XPRESS software developed by Dash Company, XA software developed by Stanford University, OSL software developed by IBM Company, CPLEX software developed by IBM Company, and Gurobi solver developed by Gurobi Optimization Company, etc. Above solving software or solver built-in algorithms included simplex algorithm, branch and bound algorithm, and cutting plane algorithm. The advantage of mathematical programming is the global optimization of the solution can be guaranteed. However, the precise solution method of global search leads to a great increase in the time and space complexity of the solution. For example, when the model contains large integer variables, the scheduling model may face the problem of dimension disaster, and the feasible solution cannot be obtained in polynomial time. Therefore, for the variable scheduling, a single mathematical programming model often cannot cover all the factors and faces problems such as large solution space and difficult calculations.

Nevertheless, the mathematical programming still plays an important role in solving the pipeline scheduling problem. For example, in natural gas pipeline scheduling, mathematical programming model can be solved directly by nonlinear optimization methods such as interior point method and Newton gradient method. In order to improve the accuracy and efficiency of solving the model under different constraints, scholars put forward many optimization methods to improve the accuracy and efficiency of solving the model. Geiler et al. [13] described how to solve the problem of using mixed integer (MIP) programming method to solve the feasibility optimization of natural gas pipeline network. The MILP model is obtained by relaxing the established MINLP model formula, which is solved by the branch and bound method of improved branch strategy. At the same time, Schmidt [14] reconstructed the problem, and an improved interior point method framework [15] is used to solve the nonlinear feasibility problem of the non-smooth complementary constraint. Cobos-zaleta et al·[16] and others used the external approximation method with equality relaxation and augmented penalty provided by solvers to solve the compressor power approximation function based on the established compressor power approximation function.

2.2.2 *Heuristic Algorithm*

The heuristic algorithm is based on intuitions or empirical constructs that give a feasible solution to the combinatorial optimization problem within an acceptable range. The method provides a quick solution or reduces the search space to reduce the complexity of the formulas used in representing real-world systems. Heuristic algorithm is a common method for solving complex optimization problems, without finding the optimal solution in polynomial time, but balances the calculation time and scheduling effect, and obtains suboptimal or satisfactory solutions with less computation. Heuristic algorithm mainly depends on domain knowledge to reduce the solving space of the problem to guide the solution, with the advantage of obtaining satisfactory solutions within acceptable computing time for large-scale problems. The heuristic rules of the heuristic algorithm can be a relatively mature abstract algorithm, such as the greedy algorithm, priority algorithm, or a set of logic rules formulated according to the specific scheduling problem and process constraints of the model.

The heuristic rules of the heuristic algorithm can be relatively mature abstract algorithms, such as greedy algorithm, priority algorithm, etc. For example, precedence algorithm is used to solve the scheduling problem of refined oil pipelines with the priority of oil demand. In this problem, each station's distribution and transportation demands may not be met at the same time, and each station has a priority order for various oil products. Therefore, the scheduling plan can be formulated by considering all stations uniformly based on the quantitative priority method. This problem can also be solved by greedy algorithm. Partial stations are preferentially processed based on the greedy algorithm, first to meet the distribution and transportation needs of the priority station, and then to meet the second priority station, etc. until the final batch is delivered. The heuristic rules of the heuristic algorithm can also be a set of solving logic rules according to specific scheduling problems and process constraints of the model. For example, Rejowski and Pinto [17] summarize a set of convergence rules for distributing operations based on the historical operation plan of actual pipelines and propose a heuristic method for optimizing distributing plans.

For the optimization of the natural gas pipeline network scheduling, the operating energy consumption of the compressor can be optimized by taking the nearest operating condition point of the compressor to the best operating condition point as the objective function. However, the energy consumption of the compressor is affected by the operating conditions of the whole pipe network and the pressure of the inlet and outlet, it is not mean that the closer it is to the best operating condition point, the smaller the energy consumption will be, and there may be different degrees of deviation under different operating conditions. In general, the result obtained by taking the minimum deviation as the objective function is not too bad, and it is usually a satisfactory solution. This is the heuristic rule for running the optimization, which can reduce the nonlinearity of the original model with the lowest energy consumption and obtain a satisfactory solution to a certain extent. The research results show that the heuristic method has a good effect on solving basic math problems such as hydraulic nonlinear scheduling problems and large-scale model solving efficiency. However, the use of this method will inevitably lead to problems such as poor optimality of the solution results and weak universality of the method.

2.2.3 *Metaheuristic Algorithm*

Metaheuristic algorithms include particle swarm optimization (PSO), ant colony optimization (ACO), genetic algorithm (GA), simulated annealing algorithm (SA), and so on.

This kind of algorithm is generally used to solve the oil product scheduling problem in stages. For example, when solving large-scale scheduling problems, the above algorithm can be used to obtain some key variables in the model, such as batch arrival time node ranking, injection station injection oil injection sequence, etc. Then the traditional mathematical programming method can be applied to solve the second stage model based on some key variables obtained. The results of the second stage can be used as the fitness function value of the metaheuristic algorithm to iterate until the model converges. Although these algorithms can solve nonlinear problems such as batch sorting, they also have the following shortcomings: (1) The solution results may converge to the local optimal solution; (2) The algorithm takes a long time to solve. If the objective function of the original problem is taken as the fitness function of the metaheuristic algorithm, the algorithm often needs to iterate hundreds or even thousands of times before convergence, and each iteration needs to complete computation of high complexity.

For the optimization of the natural gas pipeline network scheduling, the components in the natural gas pipeline network can be divided into active components and passive components. Among them, the passive component is the pipe section. The passive refers to that a pipe network composed of any pipe section, no matter what its topology or pipe network parameters are, as long as the pressure and flow of one section of the pipe network are determined, the hydraulic state at any position of the whole pipe network will be determined. The other part is active components, such as compressors, valves, etc. For active components, the system can choose to adjust the compressor speed, pressure of inlet and outlet, and other control methods to change the operating state of the system. Therefore, the initialization particle of the PSO algorithm is to determine the control parameters of two active components of the pipe network. After the initial hydraulic state is determined, the hydraulic simulation of the pipeline and the calculation of various costs are carried out, then the fitness of the particles is calculated, and finally, it is judged whether the particles have reached the maximum number of iterations. If it is reached, the result will be output, and the position and velocity of the particle need to be updated if it is not reached.

2.2.4 *Dynamic Programming Algorithm*

The optimization problem satisfying this property can be solved by dynamic programming algorithm. The scheduling optimization of the gas pipeline network can divide the process into several interrelated stages. In each stage, decisions should be made and then a complete plan with good effect can be obtained for all stages. To improve the effect of the complete plan, decisions for each stage should be determined based on current state and impacts on the future. The way of processing a problem as a chain-like structure and solving the problem by multi stages is called multi-stage decision-making process. The problem is then called multi-stage decision-making problem. In the multi-stage decision-making problem, decisions taken in each stage are generally related with time because it should be determined based on current state and impacts on the future. The final plan is generated under the changing state, so it's dynamic. This process of optimizing multi-stage decision-making problems is called the dynamic programming method.

The decision variable of the natural gas pipeline network scheduling optimization problem is usually the control value of compressor pressure or pressure ratio. In the hydraulic system of the whole pipeline, the pressure between different compressor stations meets no aftereffect; the hydraulic change condition after a compressor station is only related to the compressor station, not to the previous compressor station. Therefore, it can be solved by dynamic programming algorithm.

2.2.5 *Data-Driven Algorithm*

Some scholars [18] have introduced self-learning and other machine learning methods into the scheduling field, enriching the solving means of scheduling problems and making the research directions of scheduling problems more diversified.

The solution method based on machine learning tries to make full use of the historical operation data of oil product pipeline and changes the formulation model of oil product pipeline scheduling plan from traditional mathematical planning and theoretical solution to data-driven solution. The method based on self-learning can solve the scheduling problem of the product oil pipeline with known time nodes. Meanwhile, it can reduce the solving time of the model on the basis of learning the history operation plan of pipeline efficiently. In the metaheuristic algorithm, when the initial solution and the optimal solution differ greatly, the calculation time of the model may be too long and the global optimal solution cannot be obtained. However, in the actual operation of pipelines, it is not very different from the historical scheduling plan for a new plan to be made when the supply plan, demand plan, and initial pipeline state of a pipeline are not very different. Therefore, the pipeline historical operation plan database can be established, and the closest historical plan can be found from the database by comparing the above parameters when making a new pipeline scheduling plan, and it can be used as the initial solution of the metaheuristic algorithm. For example, in order to get the initial solution of sequencing batch arrival time node, according to the known parameters of the new plan, a similar scheduling plan can be found from the history plan database by using the method of fuzzy clustering, sorting by the time node of the plan as the initial solution of the new plan, so as to accelerate the convergence speed of the algorithm and improve calculation results [6].

For the data-driven natural gas pipeline scheduling optimization, considering the monthly, daily, and hourly gas consumption non-uniformity of off load nodes, the natural gas regulation department needs to quickly respond to the flow fluctuation and realize the accurate control of pipeline pressure, compressor start and stop, and flow distribution in the station. Based on this, it is necessary to apply the hydraulic simulation model to simulate the historical dispatch plan of the natural gas pipeline network under different supply and demand conditions, and to mine the association rules between the flow of each node and the energy consumption of the compressor. Based on the forecast of node supply and demand trends and the relationship between energy and energy consumption, a steady-state scheduling optimization model for natural gas pipeline network is established to realize rapid scheduling optimization of unified flow direction regulation and unified pressure regulation of large-area natural gas pipeline network.

2.3 Conclusion

With the development of intelligent pipelines, pipeline scheduling gradually develops from human–computer interaction to intelligence. Although the scheduling software has continuously improved the degree of automation, to formulate an efficient and flexible oil product pipeline scheduling plan, it must be based on the original research, with the help of modern technologies such as the Internet of Things, big data, artificial intelligence, etc., to achieve intelligent scheduling. This chapter summarizes existing research from the perspective of modeling methods and solution algorithms. At present, the research on modeling methods is relatively mature, and intelligent scheduling is biased toward the development of efficient and accurate solution algorithms. The prospects for intelligent pipeline scheduling are as follows:

(1) There are currently some data-driven algorithms. However, in order to improve the solution efficiency and solution quality, it is still necessary to develop new algorithms based on big data mining combined with existing algorithms. In this way, real-time decisions about pipeline scheduling can be supported.
(2) In order to realize the transition from real-time scheduling to active scheduling, it is necessary to closely combine the changes in consignment demand and the changes of pipeline operation status to achieve quantitative prediction. Efficient and accurate solution algorithms can be developed based on the prediction results and the data obtained by the full perception of pipelines.

References

1. MAGATÃO L, ARRUDA L V R, NEVES F. A mixed integer programming approach for scheduling commodities in a pipeline[J/OL]. Computers and Chemical Engineering, 2004, 28(1–2): 171–185. https://doi.org/10.1016/S0098-1354(03)00165-0.
2. HERRÁN A, DE LA CRUZ J M, DE ANDRÉS B. A mathematical model for planning transportation of multiple petroleum products in a multi-pipeline system[J/OL]. Computers and Chemical Engineering, 2010, 34(3): 401–413. https://doi.org/10.1016/j.compchemeng.2009.11.014.
3. CAFARO D C, CERDÁ J. Optimal scheduling of multiproduct pipeline systems using a non-discrete MILP formulation[J/OL]. Computers and Chemical Engineering, 2004, 28(10): 2053–2068. https://doi.org/10.1016/j.compchemeng.2004.03.010.
4. CAFARO D C, CERDÁ J. A rigorous mathematical formulation for the scheduling of tree-structure pipeline networks[J/OL]. Industrial and Engineering Chemistry Research, 2011, 50(9): 5064–5085. https://doi.org/10.1021/ie101462k.
5. ZHANG H R, LIANG Y T, XIAO Q, et al. Supply-based optimal scheduling of oil product pipelines[J/OL]. Petroleum Science, 2016, 13(2): 355–367. https://doi.org/10.1007/s12182-016-0081-x.
6. HAORAN Z, YONGTU L, QI L, et al. A self-learning approach for optimal detailed scheduling of multi-product pipeline[J/OL]. Journal of Computational and Applied Mathematics, 2018, 327: 41–63. DOI: https://doi.org/10.1016/j.cam.2017.05.040.
7. LIU Y, XIE G, CHEN X, et al. An active scheduling policy for automotive cyber-physical systems[J/OL]. Journal of Systems Architecture, 2019, 97: 208–218. https://doi.org/10.1016/j.sysarc.2018.11.004.

8. CASTRO P M, GROSSMANN I E, ZHANG Q. Expanding scope and computational challenges in process scheduling[J]. Computers & Chemical Engineering, 2018, 114(JUN.9): 14–42. https://doi.org/10.1016/j.compchemeng.2018.01.020.
9. RAMAN R, GROSSMANN I E. Modelling and computational techniques for logic based integer programming[J]. Computers & Chemical Engineering, 1994, 18(7): 563–578. https://doi.org/10.1016/0098-1354(93)E0010-7.
10. MOSTAFAEI H, CASTRO P M. Continuous-time scheduling formulation for straight pipelines[J/OL]. AIChE Journal, 2017, 63(6): 1923–1936. https://doi.org/10.1002/aic.15563.
11. CASTRO P M, MOSTAFAEI H. Product-centric continuous-time formulation for pipeline scheduling[J/OL]. Computers and Chemical Engineering, 2017. https://doi.org/10.1016/j.compchemeng.2017.04.023.
12. CASTRO P M. Optimal Scheduling of Pipeline Systems with a Resource-Task Network Continuous-Time Formulation[J]. Industrial & Engineering Chemistry Research, 2010, 49(22): 11491–11505. https://doi.org/10.1021/ie1010993.
13. GEILER B, MARTIN A, MORSI A. Chapter 6: The MILP-relaxation approach[M]. Evaluating Gas Network Capacities, 2015. https://doi.org/10.1137/1.9781611973693.ch6.
14. SCHMIDT M. A Generic Interior-Point Framework for Nonsmooth and Complementarity Constrained Nonlinear Optimization[J]. 2013. https://doi.org/10.15488/8161.
15. SCHMIDT M. An interior-point method for nonlinear optimization problems with locatable and separable nonsmoothness[J/OL]. EURO Journal on Computational Optimization, 2015, 3(4): 309–348. https://doi.org/10.1007/s13675-015-0039-6.
16. COBOS-ZALETA D, RÍOS-MERCADO R Z, UNIVERSITARIA C, et al. A MINLP Model for a Minimizing Fuel Consumption on Natural Gas Pipeline Networks Universidad Autónoma de Nuevo León[C]. 2002: 27–31.
17. REJOWSKI R, PINTO J M. Scheduling of a multiproduct pipeline system[J/OL]. Computers and Chemical Engineering, 2003, 27(8–9): 1229–1246. https://doi.org/10.1016/S0098-1354(03)00049-8.
18. LIAO Q, ZHANG H, XIA T, et al. A data-driven method for pipeline scheduling optimization[J/OL]. Chemical Engineering Research and Design, 2019, 144: 79–94. https://doi.org/10.1016/j.cherd.2019.01.017.

Chapter 3
Demand Side Management in Smart Pipeline Networks

Lin Fan and Xiao Wang

Abstract With the development of the technology and information science, nowadays it has the potential to make oil and gas supply systems more efficient and safety in terms of system operation. This chapter describes a novel model for intelligent demand side management (DSM) system, which incorporates customer demand analysis and forecasting, customer reaction analysis, dynamic pipeline network prediction, rapid supply reliability assessment, multi-objective optimization, and dynamic pricing. The DSM approach uses a dynamic pricing strategy to smooth load patterns, increase firm profit, and improve system dependability.

3.1 Development of Demand Side Management

Natural gas is a clean energy which has been used for social development and economic prosperity. Natural gas has developed rapidly in recent decades. Natural gas will contribute for 25% of global energy consumption by the mid-twentieth century, depending on the report from energy organization [1]. The efficient energy is more competitive than alternative renewable, such as nuclear energy and wind energy, due to its supply dependability. As a result, optimizing and trying to balancing the natural gas supply and demand is critical.

Various efficient management strategies have been established in various nations to successfully manage the functioning of natural gas market. For example, the natural gas market in the USA established the competitive structure, in which the market determines the natural gas price. Besides, the supply capacity of pipelines is open to different social providers and consumers [2]. As the European market leader, UK trading gas as a short-term commodity but not on long-term agreements

L. Fan (✉)
PetroChina Planning & Engineering Institute, 100083 Beijing, China
e-mail: fanlin927@126.com

X. Wang
National Engineering Laboratory for Pipeline Safety; Beijing Key Laboratory of Urban Oil and Gas Distribution Technology, China University of Petroleum, 102249 Beijing, China
e-mail: wangxiao_cup@163.com

H. Su et al. (eds.), *Advanced Intelligent Pipeline Management Technology*,
https://doi.org/10.1007/978-981-19-9899-7_3

[3]. However, the current gas market management cannot maintain supply dependability and minimize operational costs. The present management's limitations can be listed as follows: First, providers produce natural gas consistently for some time, but since end-user natural gas needs are always changing, natural gas shortages during peaks may lead to intolerable penalty prices. Furthermore, natural gas can be stored underground or in specifically built facilities to help balance the gas demand and gas supply. However, these storage installations sometimes require strict geological conditions and are constrained by the engineering design and gas source conditions. The preceding studies have emphasized the importance and benefits of operators, while customers' satisfactions and economic costs have received little consideration. It is critical to develop a flexible and intelligent decision-making skeleton to improve the system operation efficiency and reduce the shortage of natural gas.

The flexibility of energy price is vital to unlock market potential and increase efficiency, according to electric power grids [4]. Demand response (DR) improves efficiency and system reliability by including CUs [5]. Price-based and incentive-based [6]—System operators may switch off various appliances in incentive-based DR schemes to minimize CUs' energy consumption [7]. Different smart grid management frameworks have proved its ability to handle DR concerns by incorporating distributed or centralized methods to portray users' energy usage patterns, intending to increase system operation dependability and optimize system profits [8]. DR management involves utility corporations, transmission operators, and end-consumers. Consumers include electric plants, industrial and commercial customers. It is a great challenge of optimization system operation and developing pricing strategy under limited source and environmental restrictions.

RL has been used to decrease micro-grid expenses and enhance energy efficiency [9]. In Ref. [10], a forecasting model identifies environmental indeterminacies, and RL is used to investigate the ideal smart grid pricing scheme. This strategy may boost revenues and minimize energy usage. In Ref. [11], model-free reinforcement learning is adopted to improve energy management efficiency leveraging operating data. RL-based solution addresses DR management for privacy-preserving clients considering the characteristics of customers. The suggested architecture minimizes the effect of random consumer perturbations [12]. Our study explores optimum management solutions for long-distance natural gas pipelines under physical limitations.

However, one cannot instantly apply power grid DR protocols to natural gas pipeline networks due to inherent incompatibility. Natural gas that is compressible causes pressure fluctuations at nodes that are slow to propagate across pipelines. Modeling process is different from electric power networks due to the complex transient characteristics of natural gas [13]. Furthermore, since natural gas pipeline networks have tougher boundary requirements, changes in demand might result in safety problems because of pressure constraints at key nodes [14]. In order to bridge research gaps, this section suggests a novel systematic DR approach. This proposed method can provide insight into maximizing profitability, ensuring operational reliability, and reduce CU dissatisfaction, for a natural gas market with a variety of consumers and supplies.

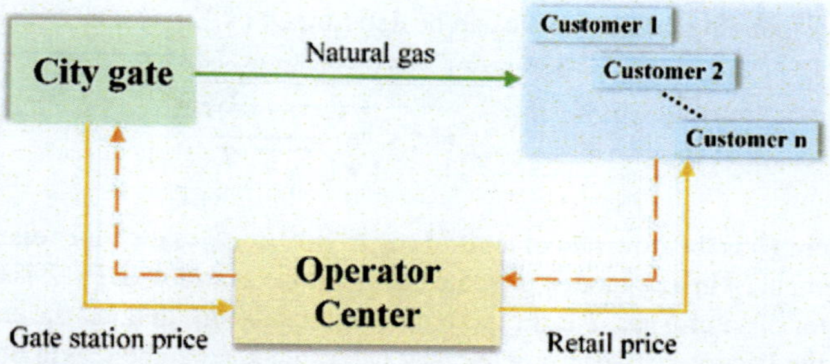

Fig. 3.1 City natural gas network model

3.2 Methodology

Forecasting of gas demand, pricing strategies developing and performance evaluation are three parts of the proposed methodology. A neural network model based on deep learning is utilized to anticipate the gas demands of consumers in the forecasting part. Then, taking into account the natural gas flow and communication between gas resources and end customers, a framework for managing the network of natural gas pipelines leveraging the advanced deep reinforcement learning is developed to address the issue of the relationship modeling between intelligent and physical models. To learn more about the suggested approach, we analyze several algorithms and assess the system's performance across a range of price ranges. In the final parts, a sensitivity analysis is carried out to take into account the influence of the incentive on the enhancement of system performance.

3.2.1 *Mathematical Models for Gas Networks*

Gate stations, CUs, and operations centers are the primary components of natural gas pipeline networks (OC). Figure 3.1 shows how the pipeline system's natural gas flow and information flow. The gate station serves as the intersection of the city pipeline networks and upstream gas sources. Natural gas prices are set by the OC in response to requests from various CUs, which are crucial for information exchange [15]. The natural gas is transported by pipelines from the gas sources to the end consumers, and compressors are constructed to increase the kinetic energy of natural gas.

3.2.2 *Pipeline Network Model*

The temperature and pressure drop along the pipeline in the delivering the natural gas downstream. Pressure, temperature, pipeline diameter, and gas propriety, the corresponding characteristic can be determined by Eq. (3.1):

$$Q_j = C_0\sqrt{\frac{(p_i^2 - p_j^2)D_{ij}}{\mu_{ij}Z\ \Delta_*\ T_{ij}L_{ij}}} \tag{3.1}$$

where Q_j is the flow rate of natural gas in Nm^3/s, D_{ij} is the inner diameter of the pipe segment *i-j* in *m*, p_i and p_j indicate the pressures at nodes, in Pa, Z is the compressible factor of natural gas. T is the gas temperature, in K, μ_{ij} denotes the energy loss caused by the friction between pipeline and natural gas, L_{ij} is the length of pipeline m–n in m, Δ_* is the specific gravity.

We use Eq. (3.2) to calculate the specific gravity:

$$\Delta_* = \frac{S_g}{S_a} \tag{3.2}$$

where S_a and S_g represent the molecular mass of gas and natural gas, respectively. The calculation of the specific gravity can be obtained by Eq. (3.3):

$$S_g = \sum S_i y_i \tag{3.3}$$

The frictional factor is calculated by the following Eq. (3.4):

$$\mu_{ij} = 0.009236\left(\frac{D_{ij}}{Q\ \Delta_*}\right)^{0.0302} \tag{3.4}$$

(1) Balance constraints

The mass input rate and outflow rate ought to be equal at every node in accordance with the mass conservation rule. The following is an expression for this restriction:

$$m_{i,\text{in}} - m_{i,\text{out}} = 0 \tag{3.5}$$

where $m_{i,\text{in}}$ and $m_{i,\text{out}}$ denote the absolute mass inflow rate and outflow rate of the ith node.

(2) Node constraints

Firstly, the node pressure has to be handled within the certain range. More specifically, the maximum permitted pressure difference of the pipeline must be higher than the distribution pressure at nodes, and the compressor input pressure should be greater than the necessary suction pressure. The pressure restriction may be expressed as:

$$p_{i,\min} < p_i < p_{i,\max} \tag{3.6}$$

where p_i is the operation pressure, in Pa; $p_{i,\min}$ denotes the minimum delivery pressure, in Pa; $p_{i,\max}$ is the maximum allowable pressure of the ith node, in Pa.

To counter the wasted energy caused by friction, gas compressors are deployed in the pipeline system. The high speed of the impeller provides kinetic energy for the passing natural gas. The value of the head may be calculated using the formula below:

$$H_{\text{pol}} = ZRT_a \frac{\sigma}{\sigma - 1} \left[\left(\frac{p_b}{p_a} \right)^{\frac{\sigma}{\sigma-1}} - 1 \right] \tag{3.7}$$

where R is the universal gas constant, T_a is the inlet temperature of gas compressor. p_b, p_a are the pressures at inlet and outlet, respectively, σ denotes the isentropic exponent. It can be derived from Eq. (3.8):

$$\sigma = \frac{\sum C_{pi} \text{My}_i}{\sum (C_{pi} \text{My}_i) - \overline{R}} \tag{3.8}$$

3.2.3 *Analysis of Customer Demands*

We assume that consumers prefer to consume natural gas when prices are low. And they prefer to minimize consumption when prices are higher than they anticipate psychologically. This serves as the basis for the management of demand response. We introduce Eq. (3.9) to explain the characteristics:

$$Q^c_{t,n} = Q^d_{t,n} \left(0.5 + \frac{1}{1 + e^{\alpha_n (\mathcal{K}_{t,n} - \phi_n) \cdot \xi_t}} \right) \tag{3.9}$$

where $\mathcal{K}_{t,n}$ denotes the gas retail price for CU n at time slot t, α_n represents the sensitivity coefficient of CU n, ξ_t is the elasticity coefficient, which describe the reaction to demand at different time periods, ϕ_n represents the psychological expected price of user n. Besides, if the retail price is equal to the psychological expected price, the supplied gas is equal to the demanded gas. It indicates that the CU will choose to adjust their demand as anticipated and that its demand is unaffected by the retail price.

3.2.3.1 Measure for Demand Fluctuation

The probability of a natural gas shortage and system performance will rise with abrupt changes in flow rates and demand. The optimization technique takes into account the gas consumption variation rate, which is strongly connected to system dependability.

In Eq. (3.10), the fluctuation of natural gas for customer n is computed as follows:

$$\varepsilon_{t,n} = (Q^c_{t,n} - \overline{Q}_n)^2 \tag{3.10}$$

where $\overline{Q_n}$ represents the average gas consumption of CU n for T period.

3.2.3.2 Definition of Customer Dissatisfaction

As the actual demand of the customer is not met, i.e., there is an imbalance between the actual demand for gas and the actual supply of gas. This supply imbalance is reflected in customer dissatisfaction. The dissatisfaction is calculated by Eqs. (3.11)–(3.13):

$$\chi_{t,n} = \beta_n Q^d_{t,n} \left(\frac{Q^c_{t,n}}{Q^d_{t,n}} \right)^{\alpha_n} - 1 \tag{3.11}$$

$$\alpha_n = \frac{1}{k_n} + 1 \tag{3.12}$$

$$\beta_n = -\frac{4}{\alpha_n} \tag{3.13}$$

where k_n represents the elasticity factor. For this specific situation, the elasticity coefficient represents the responsiveness and elasticity of the quantity demanded of service to a change in its price. Generally, the value of the elasticity factor is negative. The critical parameters α_n and β_n are associated with k_n.

3.2.4 Operating Center Profit Model

The gas sources deliver the natural gas to different end-consumers at each time slot. According to OC, the profits are based on the volume of natural gas sold downstream. We assume that the gas price at gate stations is constant, which makes it more clearly illustrate how retail pricing influences CU's demands. Equation (3.14) may be used to compute OC's profit.

$$\delta_{t,n} = Q^c_{t,n}(\kappa_t - \vartheta) \tag{3.14}$$

3.2.5 Objective Functions

In this section, we characterize the reward function of the optimization problem by using objective function to improve system performance, which taking into account a variety of factors, including OC profit, CU discontent, and demand variability [16] using Eq. (3.15):

$$R(\boldsymbol{s}_t, \boldsymbol{a}_t, \boldsymbol{s}_{t+1}) = \sum_{n=1}^{N} \sum_{t=1}^{T} C_p \cdot \delta_{t,n} - C_c \cdot \chi_{t,n} - C_b \cdot \varepsilon_{t,n} \tag{3.15}$$

where C_p denotes the utility of OC's profit through getting one unit, C_c denotes the dissatisfaction spending of one unit, C_b denotes the spending of fluctuation of the demand. C_p, C_c, and C_b are set properly which indicate monetary unit (m.u.) per-unit of natural gas.

3.2.6 Intelligent Algorithm for Gas Pipeline System Management

An intelligent algorithm that shows promise is reinforcement learning. The sequential decision-making issues are successfully applied in numerous fields which is largely attributed to its flexibility on complicated nonlinear situations [17]. Reinforcement learning does not need extensive previous information, unlike conventional heuristic algorithms like the particle swarm optimization, ANN, and greedy algorithm, and it may learn and adapt constantly in the dynamic environment. It is significant for unpredictable and adaptable natural gas markets to make sensible decisions. Agent, environment state *s*, action *a,* and reward *r* represent four basic elements in typical reinforcement learning. Generally speaking, the deep learning methods are as follows: To begin with, the state matrix is passed to the agent at time slot *t*; what's more, the agent due to its policy $p_\pi(a|s)$ select an action $a_t \in \mathcal{A}$ and carry out the action which based on the environment; the environment follows the transfer probability $p(s_{t+1}|s_t)$, and then state s_t becomes to a new state s_{t+1}; in the back of receiving a scalar reward, which defines how good the action has been selected, the agent through maximizing the expected total reward to update its strategy, which will be token over from now on.

3.2.6.1 Solving the Pricing Problem

In order to deal with the dynamic pricing problem in natural gas pipeline networks, different CUs must be determined by the agent OC at different times. It can be expressed as an extensive Markov decision process (MDP) [18]. The key elements include: discrete time t, action $\boldsymbol{a}_t(\kappa_{t,n})$, state $\boldsymbol{s}_t(Q^c_{t,n}, Q^d_{t,n})$, reward $R(r(Q^c_{t,n}|Q^d_{t,n}, \kappa_{t,n}))$, and combined state probability $T(p(\boldsymbol{s}_{t+1}|\boldsymbol{s}_t, \boldsymbol{a}_t))$. In order to deal with MDP, finding a way that delivers big returns is important; v_* refers the optimal function with state value:

$$v_*(s) = \max_{\pi} v_\pi(s) = \max \mathbb{E}_\pi \left[\sum_{k=0}^{\infty} \gamma^k R_{t+k+1} | S_t = s \right] \tag{3.16}$$

where the policy π denotes a mapping state transitions, the worth of state s under a policy π, referred $v_\pi(s)$, is the returns when beginning with in s and following π. After that, $\mathbb{E}_\pi[\cdot]$ denotes the average value under the stochastic transitions, and $\gamma \in [0, 1]$ [0, 1] denotes a discount factor.

In the same way, the state-action value function can be determined $Q : S \times A \to \mathbb{R}$ due to Eq. (3.17) as below:

$$Q^*(\boldsymbol{s}, \boldsymbol{a}) = \mathbb{E}[R_{t+1} + \gamma v_*(\boldsymbol{s}_{t+1}, \boldsymbol{a})|S_t = \boldsymbol{s}_t, A_t = \boldsymbol{a}_t] \tag{3.17}$$

3.2.6.2 Dynamic Pricing Method Based on Deep Q-Learning (DQL)

Decision-making issues have been resolved in a variety of contexts using Q-learning (QL). The classic QL uses a table with rows and columns to describe the action and state spaces as a value-based approach. The values of the matching values are kept in the cells of this table in each row. Since the spaces may be quite large and the cost and complexity will be unacceptably high, using QL to solve a genuine, issue on the large scale can be challenging. The framework of combining DQL with DSM for pipeline system is depicted in Fig. 3.2.

Through replacing value function from the standard QL algorithm to a deep Q-learning (DQL) with parameters θ, which can be described as $Q(s, a, \theta) \approx Q^\pi(s, a)$. To decrease the uncertainty produced by data relationship, a parameter vector θ^- is used here. The parameter θ^- denotes the self-updating function. The objective function of gradient descent is as follows in Eq. (3.18):

$$l(\theta) = \mathbb{E}[(\boldsymbol{r} + \gamma \max_{\boldsymbol{a}_{t+1}} Q(\boldsymbol{s}_{t+1}, \boldsymbol{a}_{t+1}, \theta^-) - Q(\boldsymbol{s}_t, \boldsymbol{a}_t, \theta))^2] \tag{3.18}$$

3.2.7 *Forecasting Model of Customer Demand*

Current approaches to energy demand forecasting include time series models, neural network models, regression models, and hybrid models. Among them, time series models have the advantages of not requiring a priori knowledge and being simple and

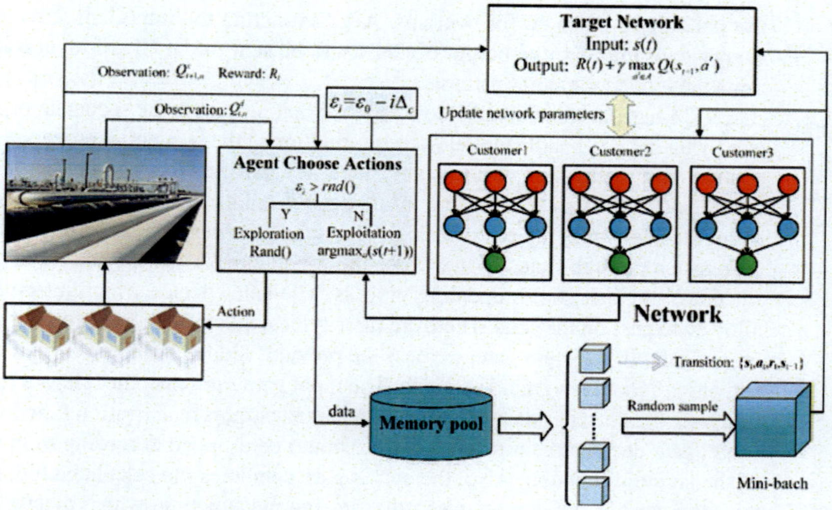

Fig. 3.2 Framework of DQL for the pipeline system

easy to operate and have been applied to energy demand forecasting for various event horizons. Similar to event series models, regression models are also widely used for energy demand forecasting due to their simplicity and promising forecasting results.

The CNN-LSTM method can extract complex features from multiple variables collected by the local sensors. It has gained popularity in spatio-temporal energy consumption forecasting problems. Here, we build a series connection to integrate their feature extraction with the forecasting model. A typical CNN network contains a convolutional layer, a pooling layer, a ReLU layer, and fully connected layer. Firstly, the convolutional layer applied convolutional operation to the input variables to extract the basic feature. Then the output of the convolutional layer will be sent to the ReLU layer, which creates a nonlinear function to learn the complex time series. The pooling layer focuses on feature integration that reduces the dimension of features and uses fewer parameters; it can reduce the complexity of computation without distortions. To improve the accuracy and generalization of the forecasting model, the aforementioned layers can be stacked to make the model deeper. Finally, the output layer extracts features to LSTM. Here, we define $x_j^0 = \{x_1, x_2, x_3, \cdots x_n\}$ as the input vector of the first convolutional layer; the number of n is the normalized data per window with length T.

The other advanced technique adopted in the convolutional network is using a pooling layer to reduce the number of parameters and computation cost in the network. It provides an approach to downsampling feature maps by summarizing the presence of features in patches of the feature map. The most common pooling operation is max-pooling; it selects the maximum value of each patch of feature map. The stride is another parameter that modifies the amount of movement over the vector. If the stride of pooling does not perfectly fit the kernel size and input size, we will use padding to make up for overlaps. After extracting the important features from energy consumption, the output of CNN will be sent to the LTSM layer to further mine the hidden temporal patterns.

LSTM has the same control flow as RNN that processes the sequence of vectors one by one; it can handle the relevant information in the sequential process by propagation. But, if the time series is very long, it's hard for RNN to predict accuracy for its poor ability in transferring information from earlier time steps to later time steps. Besides, the gradient vanishing is another problem as the information back propagates through time. To overcome these bottlenecks, LSTM uses several gates to control the cell state's updating process. The gates decide which information is allowed to pass on the cells. There are three typical gates: input gate, forget gate, and output gate. In the input gate, we pass the previous hidden state and current input to activation function sigmoid and tanh. The input gate can select the relevant information from the current step and integrate regulated outputs of activation functions. The forget gate decides the information that should be dropped according to the output of the sigmoid function. Also, the output gate combines the calculated hidden state with the current state and passes it to the sigmoid function, which determines the next hidden state.

To extract features from the multivariate time series collected from different locations, we integrate the spatial and temporal modeling capability of the CNN and LSTM model. To be more specific, we use the CNN model as the filter to explore the spatial relationship between adjacent locations; after that, we use LSTM layers to extract the temporal dependencies of features extracted from the CNN model. The structure of the CNN-LSTM model can be seen in Fig. 3.3.

As shown in Fig. 3.3, the CNN-LSTM model contains a convolutional layer, pooling layer, flatten layer, LSTM layer, and dense layer. The input time series

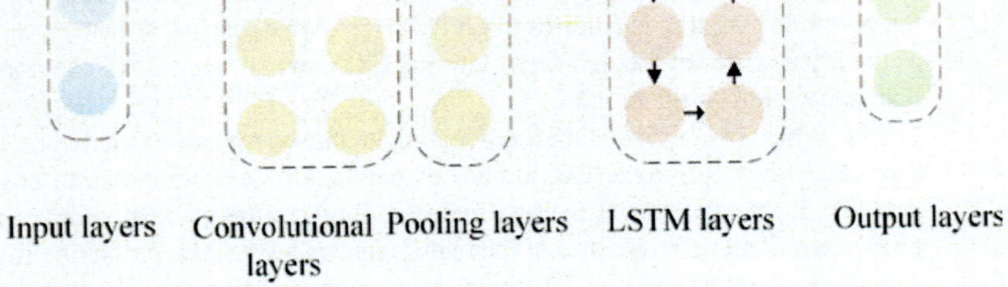

Fig. 3.3 Structure of the CNN-LSTM model

passes the stacking layers in sequence and outputs the predicted energy consumption in the last layers. At first, the units of the convolutional layer locally connect to the input layer; the kernels share the same weight matrix. The sparse connectivity and parameter sharing decrease the model parameter and simplify the training process than compromising accuracy. Besides, the pooling layer is used to downsample the output features, considering its advantages in dimensional reduction and overfitting prevention. We use the max-pooling technique to summarize the important local features. The flatten layer integrates feature maps and converts these features into a one-dimensional vector. Finally, the memory block of LSTM receives the extracted features and outputs the forecasting results in the dense layer.

It should be noted that to evaluate the performances of the forecasting model in different horizons, the number of neurons of the dense layer should be matched with various horizon times. The energy consumption forecasting may evolve into sequence-to-sequence forecasting problems.

To compare the forecasting performances of different models comprehensively, we use three criteria here, including mean absolute error (MAE), mean absolute percentage error (MAPE), and root mean squared error (RMSE), as in Eqs. (3.19)–(3.21):

$$\mathrm{MAE} = \frac{1}{N}\sum_{i=1}^{N}\left|P_i^f - P_i^a\right| \tag{3.19}$$

$$\mathrm{MAPE} = \frac{1}{N}\sum_{i=1}^{N}\left|\frac{P_i^f - P_i^a}{P_i^a}\right| \tag{3.20}$$

$$\mathrm{RMSE} = \sqrt{\frac{1}{N}\sum_{i=1}^{N}(P_i^f - P_i^a)^2} \tag{3.21}$$

where P_i^a represents the actual energy consumption at time slot i, and P_i^f denotes the forecasted energy consumption at time slot i.

3.3 Case Study

Figure 3.4 depicts a natural gas pipeline network that consists of two supplies, 16 pipelines, two compressor stations, three demand sites, and pressure ratios of between 1.02 and 1.18.

Forecasting the natural gas consumption is essential for efficient management for pipeline system. The predictions model's input in this work is the gas demand at time t. The neural network serves as the foundation for the natural gas forecasting. Following a "50–30–20%" division to split the historical data. N The number of the hidden layer ranges from 2 to 4, and the hidden unit sizes can range from [100 200 300

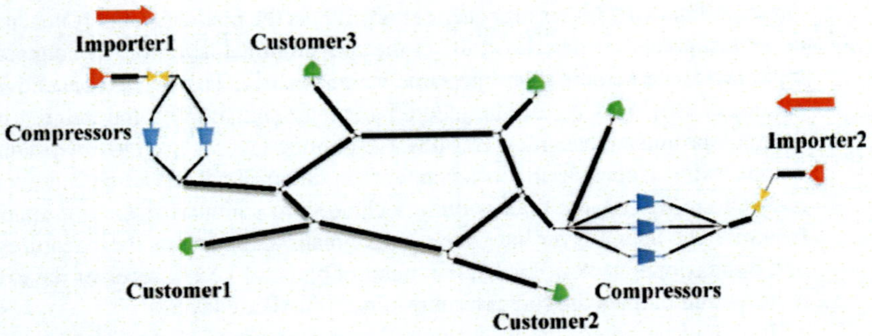

Fig. 3.4 Structure of natural gas pipeline

400]. The bi-LSTM layer that comes after an LSTM layer, with hidden unit counts of 300 and 200, respectively, turns out to be the best hidden layer architecture. The forecasting results are analyzed in detailed by comparing the forecasting performance of different models.

Figure 3.5 shows the outcomes of fixed price and dynamic pricing. We can observe that, in the case of dynamic pricing, the retail prices are correlated with the varying patterns in the demand for natural gas. This may be explained by the fact that when CU's gas requirement is greater than usual, the agent (OC) prefers to improve the retail price to reduce the gas consumption. On the other hand, when in the consumption valley, the OC prefers to decrease the retail price, to attract consumers to improve their gas consumption. The peak shifting is clearly impacted by the dynamic pricing scheme.

As shown in Fig. 3.6 by the rise in OC's profit and CU's satisfaction, it can be shown that dynamic pricing outperforms fixed pricing. According to the findings, the suggested algorithm may increase OC profit and lower demand fluctuation without compromising CU satisfaction (Table 3.1).

3.4 Conclusion

In this chapter, we provide a brand-new, methodical approach for making decisions on the operation of natural gas pipeline networks. To thoroughly optimize system performances, taking into account OC profit, CU unhappiness, and operation dependability, the suggested management framework coupled with RL is established. To anticipate the demand for natural gas, a customized forecasting model is created. The benefits of the pricing framework suggested rest on both creating tactics in line with CU patterns and the dynamically effective improvement brought about by intelligent algorithm. The findings demonstrate that the dynamic pricing method does indeed perform better than fixed pricing in raising CU satisfaction and encouraging peak shaving while saving money. For the average action-state value, we have compared

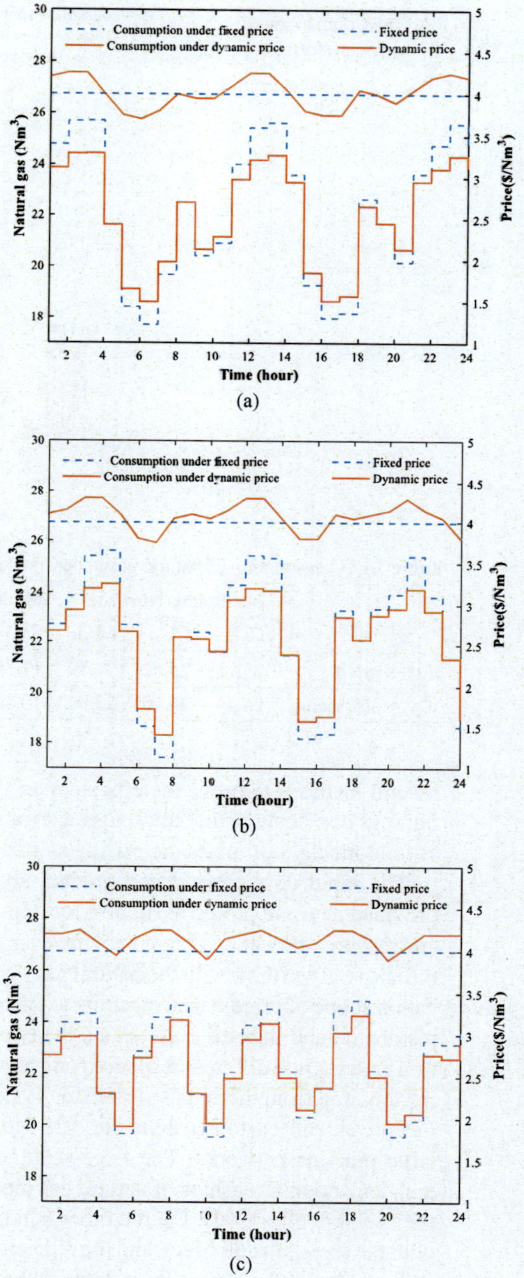

Fig. 3.5 Pricing results comparison for customers

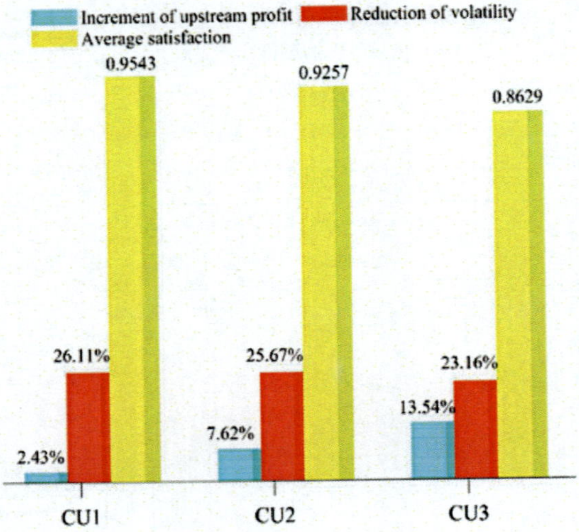

Fig. 3.6 Comparison of customers' performance

Table 3.1 Consumption of natural gas under fixed and dynamic pricing schemes

	Maximum (Nm^3/h)			Minimum (Nm^3/h)			Average (Nm^3/h)		
	CU1	CU2	CU3	CU1	CU2	CU3	CU1	CU2	CU3
Fixed pricing	25.71	25.62	24.89	17.73	17.43	19.56	22.30	22.39	22.54
Dynamic pricing	24.44	24.40	23.94	18.40	18.30	20.06	22.03	22.20	22.40

several methods to prove the efficiency of the proposed method. The comparison's findings demonstrate that the suggested framework exceeds existing algorithms and asymptotically optimizes the pricing strategy.

The rapid development and gradual application of intelligent algorithms have provided new and powerful support for improving the reliability of system operation and resource allocation efficiency. However, there are still two problems in combining the above algorithms with the natural gas system: (1) the application target and application scope of price-based incentives. The main consumers of natural gas are urban residential and industrial users, and the two types of users have different consumption patterns and different response patterns to prices. The modeling must incorporate information from the actual production system and analyze the relationship between incentives and customer behavior. (2) Efficiency of algorithm implementation in large pipeline networks. The large-scale, complex structure and changing supply and demand relationships of natural gas pipeline systems require short time to solve the problem effectively. On the other hand, the complexity of the algorithm grows with the size of the problem, and the efficiency of the solution decreases significantly, so the value of the algorithm in large network systems needs to be further explored and verified.

References

1. Lu W, Su M, Fath BD, Zhang M, Hao Y. A systematic method of evaluation of the Chinese natural gas supply security. Applied Energy 2016;165:858–67. https://doi.org/10.1016/j.apenergy.2015.12.120.
2. Huo J. Comparing the natural gas pipeline pricing between Europe and America and the revelation to China. Energy Procedia 2011;5:659–63. https://doi.org/10.1016/j.egypro.2011.03.116.
3. van Goor H, Scholtens B. Modeling natural gas price volatility: The case of the UK gas market. Energy 2014;72:126–34. https://doi.org/10.1016/j.energy.2014.05.016.
4. Shehzad Hassan MA, Chen M, Lin H, Ahmed MH, Khan MZ, Chughtai GR. Optimization modeling for dynamic price based demand response in microgrids. Journal of Cleaner Production 2019;222:231–41. https://doi.org/10.1016/j.jclepro.2019.03.082.
5. Hamwi M, Lizarralde I, Legardeur J. Demand response business model canvas: A tool for flexibility creation in the electricity markets. Journal of Cleaner Production 2021;282. https://doi.org/10.1016/j.jclepro.2020.124539.
6. Lu X, Zhou K, Zhang X, Yang S. A systematic review of supply and demand side optimal load scheduling in a smart grid environment. Journal of Cleaner Production 2018;203:757–68. https://doi.org/10.1016/j.jclepro.2018.08.301.
7. Vázquez-Canteli JR, Nagy Z. Reinforcement learning for demand response: A review of algorithms and modeling techniques. Applied Energy 2019;235:1072–89. https://doi.org/10.1016/j.apenergy.2018.11.002.
8. Irtija N, Sangoleye F, Tsiropoulou EE. Contract-theoretic demand response management in smart grid systems. IEEE Access 2020;8:184976–87. https://doi.org/10.1109/ACCESS.2020.3030195.
9. Kim BG, Zhang Y, Van Der Schaar M, Lee JW. Dynamic pricing and energy consumption scheduling with reinforcement learning. IEEE Transactions on Smart Grid 2016;7:2187–98. https://doi.org/10.1109/TSG.2015.2495145.
10. Wen L, Zhou K, Li J, Wang S. Modified deep learning and reinforcement learning for an incentive-based demand response model. Energy 2020;205. https://doi.org/10.1016/j.energy.2020.118019.
11. Tan Z, Zhang X, Xie B, Wang D, Liu B, Yu T. Fast learning optimiser for real-time optimal energy management of a grid-connected microgrid. IET Generation, Transmission & Distribution 2018;12:2977–87. https://doi.org/10.1049/iet-gtd.2017.1983.
12. Ghasemkhani A, Yang L, Zhang J. Learning-Based Demand Response for Privacy-Preserving Users. IEEE Transactions on Industrial Informatics 2019;15:4988–98. https://doi.org/10.1109/tii.2019.2898462.
13. Durgut İ, Leblebicioğlu MK. State estimation of transient flow in gas pipelines by a Kalman filter-based estimator. Journal of Natural Gas Science and Engineering 2016;35:189–96. https://doi.org/10.1016/j.jngse.2016.08.062.
14. Wu X, Li C, Jia W, He Y. Optimal operation of trunk natural gas pipelines via an inertia-adaptive particle swarm optimization algorithm. Journal of Natural Gas Science and Engineering 2014;21:10–8. https://doi.org/10.1016/j.jngse.2014.07.028.
15. Paltsev S, Zhang D. Natural gas pricing reform in China: Getting closer to a market system? Energy Policy 2015;86:43–56. https://doi.org/10.1016/j.enpol.2015.06.027.
16. Rocchetta R, Bellani L, Compare M, Zio E, Patelli E. A reinforcement learning framework for optimal operation and maintenance of power grids. Applied Energy 2019;241:291–301. https://doi.org/10.1016/j.apenergy.2019.03.027.
17. Cancela H, Murray L, Rubino G. Efficient Estimation of Stochastic Flow Network Reliability. IEEE Transactions on Reliability 2019;68:954–70. https://doi.org/10.1109/tr.2019.2897322.
18. Poolla C, Ishihara AK, Milito R. Designing near-optimal policies for energy management in a stochastic environment. Applied Energy 2019;242:1725–37. https://doi.org/10.1016/j.apenergy.2019.01.228.

Chapter 4
Life Cycle Analysis of Pipeline Networks

Guotao Wang and Liqiao Huang

Abstract As climate issues gradually attract public attention worldwide, the operation and construction of pipeline networks have been attached with new energy-saving and emission-reduction targets. This chapter systematically sorts out and summarizes the carbon emission sources and environmental impacts of pipelines during the whole life cycle. The models for the life cycle assessment are also clarified from the perspective of construction, operation, fugacity, leakage, recycle, and disposal. Through the result analysis from the existing work, this chapter identifies the driving factors of emissions, including pipe length, diameter, and throughput. Finally, relevant recommendations are put forward based on the results to help reduce emissions from pipe transportation.

4.1 Environmental Impacts of Pipelines

Researches on energy conversion to reduce carbon emissions are attractive to more and more countries due to the significant impact of carbon emissions on climate change. Unfortunately, fossil fuels still dominate the fuel market as their wide applications and irreplaceable roles in heavy transportation. As approximately 24% of carbon dioxide emissions come from transportation sector, the consumption of fossil fuels is the core factor affecting global warming. There is 3.92×106 kt oil which

G. Wang (✉) · L. Huang
Beijing Key Laboratory of Urban Oil and Gas Distribution Technology, China University of Petroleum-Beijing, Fuxue Road No.18, Changping, China
e-mail: loganwang6@163.com

L. Huang
e-mail: 549524447@qq.com

G. Wang
Department of Building Environment and Energy Engineering, The Hong Kong Polytechnic University, Hong Kong SAR, China

L. Huang
Graduate School of Engineering, University of Tokyo, Tokyo, Japan

H. Su et al. (eds.), *Advanced Intelligent Pipeline Management Technology*,
https://doi.org/10.1007/978-981-19-9899-7_4

has been consumed in 2018, and the oil consumption is predicted to keep growing. At the same time, the global demand for natural gas has increased dramatically in the past two decades, indicating a rapid increase in carbon emissions from natural gas. Both evaluating fuel efficiency and choosing future alternative fuels need the awareness of carbon emissions of various fuels.

To comprehensively evaluate carbon emissions and energy consumption, not only the end-use should be considered, but also its production, transportation, or disposal needs to be taken into account. As pipelines remain the most economical and widespread mode for inland and near-shore oil and gas transportation, pipeline networks worldwide have been expanding rapidly. To quantify the environmental effects of transporting oil products and natural gas, detailed emission assessment techniques for oil products and natural gas pipeline networks are necessary. Especially, pipeline transportation undertakes most transmission tasks for delivering oil products and natural gas, which is doubtlessly a key area of environmental effects analysis.

4.2 Carbon Emission Stages of Pipeline Transport System

Life cycle assessment (LCA) is extensively applied to quantitatively calculate carbon emissions from different activities, covering stages of "cradle to grave" of assessment objectives. For this section that concentrates on transportation system of refined oil and natural gas, emission stages in the life cycle of oil products and natural gas pipelines can be divided into four parts, including construction, operation, fugitiveness, and recycling.

Construction of pipe segments and other infrastructures is necessary before operating it officially. Usually, long-distance pipelines no matter transporting oil or natural gas consist of intermediate stations and the pipe body. Intermediate stations are arranged along the pipeline, which include pump stations, valve chambers, and other auxiliary facilities. Pump stations relying on the work of pumps are applied to provide power for the pipeline transmission, which is the most energy-consuming part of pipe operation. Correspondingly, enormous energy consumption in pump stations also means great carbon emissions caused by fossil fuel combustion. Compared with oil products pipelines, carbon emissions of pipelines transporting natural gas more depend on pump stations. Here, the production and processing of steel, the forging of materials, and its transportation to the corresponding location consist of activities related to pipeline construction. Soil excavation, pipeline laying, soil backfilling, and road compaction are indispensable because pipelines are always buried underground in order to avoid terrain restrictions and reduce floor space. Besides, anticorrosive layer needs to cover pipelines underground for protecting pipelines. Fig. 4.1a concludes activities in the pipeline construction stage.

The transmission process has a significant impact on oil products pipeline due to which multi-products are transported in pipeline simultaneously, and this process is called batch transportation in petroleum industry. Oil mixture, namely contamination

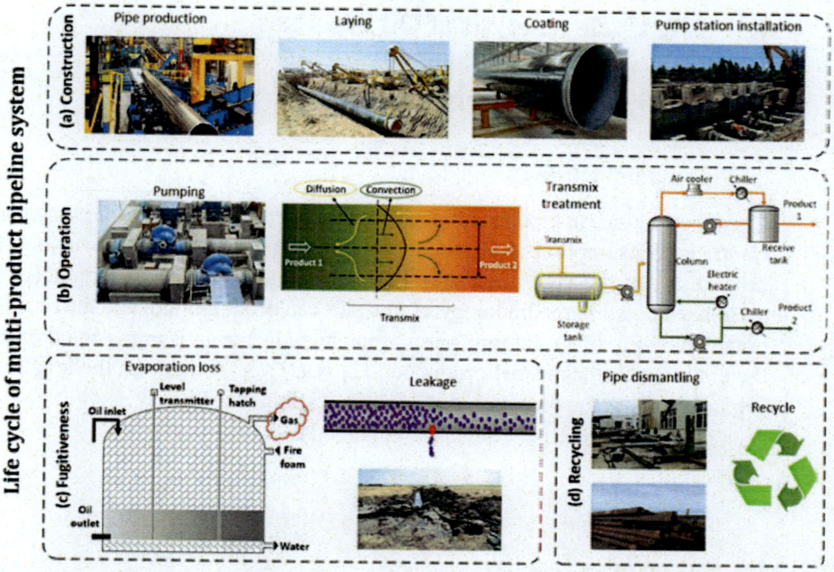

Fig. 4.1 Different emission phases in the life cycle of oil products pipeline system

or transmix, will be produced in batch transportation, as shown in Fig. 4.1b. In the practical engineering of pipeline operation, objectives containing maximizing the demand of consumers and minimizing operational costs should be considered as well. For oil products pipelines, operators need to make scheduling plans according to the capacity of upstream refineries and demand of downstream market to determine the injection plan effectively. As for natural gas pipelines, operation plan is essential for reducing energy consumption and carbon emissions.

Figure 4.1c shows fugitive emissions existing in pipeline systems, but some the emissions are designed in purpose. For instance, the venting of oil tanks has ability of avoiding overpressure. In addition, leakage caused by pipe corrosion, natural disasters, and other human factors will lead to both economic loss and environmental pollution. Leakage still happens in sometimes although protection measures like anticorrosion coating have been implemented. After years of operating, a pipeline will be declared obsolete eventually, and this is the time for recycling it, or metals and other recyclable materials would be discarded and cause irreversible pollution to the soil. Thus, recycling stage is worth of considering as indicated in Fig. 4.1d

4.3 Life Cycle Analysis of Oil Pipelines

4.3.1 Infrastructure Construction

(a) Pipe manufacturing

Carbon emissions in pipe manufacturing mainly come from fossil fuels combustion, electricity consumption, and non-energy-related emissions [1]. As the manufacturing process is not the core content of computing emissions of oil products pipeline system, calculation methodology of this part can be simplified, and the result can be obtained by Eq. (4.1). Emission from pipe manufacturing is represented by E_1, and the emission factor of steel production [2] is EF_{steel}. L_i denotes the length of pipe segment i, D_i is the diameter, and δ_i is the wall thickness.

$$E_1 = \sum_{i=1}^{n} \pi\left((D_i + \delta_i)^2 - \delta_i^2\right) L_i EF_{\text{steel}} \tag{4.1}$$

(b) Anticorrosion coatings

To protect pipelines for oil corrosion and stray current corrosion, anticorrosion coatings and cathodic protection are adopted. Long-distance pipelines usually use coal char paint, polyethylene (PE) double-layer structure, PE three-layer structure or fusion bonded epoxy (FBE) as anticorrosion coatings [3]. Here, we select 3PE coatings for carbon emissions calculation, and the emission from anticorrosion coatings E_2 is computed by Eq. (4.2):

$$E_2 = \sum_{i=1}^{n} \pi\left((D_i + \delta_i + w_i)^2 - (D_i + \delta_i)^2\right) L_i EF_{\text{coating}} \tag{4.2}$$

where EF_{coating} is the emission factor of coating materials obtained through Ecoinvent 3 database, w_i indicates the corresponding thickness of coating of the i-th pipe section.

(c) Pipe transportation and laying

For buried laying, trench excavation is often used, and excavators are required to complete the construction. Considering the large differences in pipeline geological conditions, the resulting emissions cannot be considered uniformly. Therefore, we assume the excavation process is adopted and corresponding factors can be estimated [4]. The width B_i of groove bottom is obtained from Eq. (4.3), The volume V_p of the soil excavated by the pipeline is obtained from Eq. (4.4). According to the specified parameters of the excavator, we can obtain the power consumption P_e in the process of soil excavation and backfill construction as shown in Eq. (4.5).

$$B_i = (D_i + 2\delta_i) + b \tag{4.3}$$

$$V_p = \sum_{i=1}^{n} B_i L_i h \tag{4.4}$$

$$P_e = W_e T_e V_p / V \tag{4.5}$$

T_e is the average working cycle of an excavation including full bucket lifting, empty bucket rotation, etc., W_e is its average power, and V_e is the standard bucket capacity.

Road transportation is common in transporting pipes and other auxiliary facilities, so the raw material processing plant is assumed to locate in the middle of the pipeline, then the average haul distance equals to a quarter of the total pipe length. Based on these, the emission E_3 of pipe laying can be calculated by Eq. (4.6), wherein EF_{elec} and EF_r are the emission factor of power grid and road transportation [5].

$$E_3 = P_e EF_{\text{elec}} + \sum_{i=1}^{n} \pi\left((D_i + \delta_i)^2 - \delta_i^2\right) L_i EF_r \tag{4.6}$$

(d) Construction of pump stations

In previous studies, the life cycle of liquid pumps from production to product delivery has been intensively studied [6, 7]. Therefore, we assume that the discharge of the production pump set is fixed. Considering the different emissions produced by pumps with different powers, we made a unified assumption for the power of the main and branch line pumps. Thus, the total emissions from the construction of the pumping station can be estimated using the Eq. (4.7). Among them, E_4 represents the emission from the construction of the pumping station, $E_{\text{pump},t}$ is the emission of a pump unit in the trunk line, n_t is the total number of pumps along the trunk line, $E_{\text{pump},b}$ is the output of a pump unit in the branch line, and n_b is the total emission of branch pumps.

$$E_4 = E_{\text{pump},t} n_t + E_{\text{pump},b} n_b \tag{4.7}$$

4.3.2 Pipe Operation

As we discussed above that the operation of oil product pipelines needs scheduling plan as the guideline. The scheduling plan can be attained by mixed integer linear programing models [8, 9], and it also helps to compute the energy consumption of pipelines. Equation (4.8) is constructed to calculate Reynolds number Re, wherein ρ_m is the density of product transported in pipe i, v_i represents the flow rate, and μ_m is the viscosity of product m. Then, flow pattern of oil in pipelines can be determined according to its Reynolds number.

$$\mathrm{Re}_i = \rho_m v_i d_i / \mu_m \tag{4.8}$$

After calculating the Reynolds number, the friction resistance coefficient related to different flow patterns can be calculated, as shown in Eq. (4.10). Then, the head loss H can be given by Eq. (4.9). In general, local friction h_ξ is equal to 1–2% of friction loss, so we use Eq. (4.11) to calculate the output power of pump N. Q is the flow rate of pipelines, and η denotes the efficiency of pump units. Finally, carbon emissions E_5 producing by energy consumption in pump stations are computed by Eq. (4.12).

$$H = h_f + h_\xi + Z_m - Z_c \tag{4.9}$$

$$h_f = \lambda \frac{L}{D} \frac{v^2}{2g} \tag{4.10}$$

$$N = \frac{\rho_h g H Q}{1000\eta} \tag{4.11}$$

$$E_5 = EF_{\mathrm{elec}} N \tag{4.12}$$

The amount of contamination is necessary for assessing the potential emission from contamination treatment procedures, and we choose the Austin and Palfrey model to compute it. Here, we assume that delivery temperature is 10 °C and contamination consists of 50% of both forward and trailing products [10–12], and the specific formulas are shown as follows:

$$lglg\left(v \times 10^6 + 0.89\right) = 0.5lglg\left(v_A \times 10^6 + 0.89\right) + 0.5lglg\left(v_B \times 10^6 + 0.89\right) \tag{4.13}$$

$$C = 11.75(L \times D)^{0.5} \mathrm{Re}^{-0.1} \left(\mathrm{Re} > \mathrm{Re}_{ij}\right) \tag{4.14}$$

$$C = 18385(L \times D)^{0.5} \mathrm{Re}^{-0.9} \times \exp\left(2.18 D^{0.5}\right)\left(\mathrm{Re} < \mathrm{Re}_{ij}\right) \tag{4.15}$$

The kinematic viscosity of product A, product B, and contamination is represented by v_A, v_B, and v. The length of contamination is represented by C and Re_{ij} refers the critical Reynolds number.

Usually, there are two ways for processing contamination: blending treatment and sent contamination to refineries. As the blend process mainly completes in tanks, energy consumption of contamination pumps contributes most carbon emissions in this process. For contamination sent to refineries, carbon emissions from energy consumption of feed pumps and heating devices are considered. Based on this, the total emissions E_6 of contamination treatment can be expressed by Eq. (4.16).

$$E_6 = EF_{\mathrm{elec}} W_{\mathrm{transmix}} + EF_{\mathrm{fuel}} C_{\mathrm{fuel}} \tag{4.16}$$

The power consumed by pumps is represented by W_{transmix}, which the compute method has been introduced above. EF_{fuel} is the emission factor of heating devices, and the fuel consumption of heating device is C_{fuel}.

4.3.3 Fugitive Emissions

Fugitive emissions including emissions of evaporation and accidental releases are inevitable produced during the processes of venting, flaring, filling, and unloading activities, and it can be obtained by existing results in leakage database and IPCC2006 guidelines [13, 14]. The life span of pipelines is assumed to be twenty years, and emission factors related to fugitive emissions of pipelines operation are verified based on the guideline. The fugitive emission E_7 is expressed by Eq. (4.17).

$$E_7 = EF_{\text{fe}_1} Q_a + EF_{\text{fe}_2} Q_a G_{\text{CH}_4} \tag{4.17}$$

Here, the fugitive emission factor of CO_2 and methane in oil pipe transportation is indicated by EF_{fe_1} and EF_{fe_2}, Q_a is the designed annual volume of transportation, and G_{CH_4} is the global warming potential of methane on the basis of CO_2 [15].

4.3.4 Recycle and Disposal

The recycle and disposal of pipe materials are required for protecting environment and reducing costs. There are many feasible proposals for recycling pipe materials, but haven't been widely applied in practical engineering [16, 17]. Thus, we proposed several assumptions based on the characteristics of the oil products pipeline, including pipe cleansing, removal, and material recycling.

1. Residue cleansing

Before cutting and removing pipeline, residual oil in the pipeline should be processed at first. Both cleansing process and cleanliness requirements of waste pipeline are different with pigging of in-service pipes. In general, a combination of auxiliary and sealing pigging is used to clean waste pipeline, wherein auxiliary pigging aims at removing solid deposits and sealing pigging is set up to treat liquid. Here, water is assumed to be the media of pipe cleansing, and the power consumption leads to the main carbon emission of this process. Thus, the calculation of carbon emissions can adopt Eqs. (4.8)–(4.12).

2. Pipe removal technologies

Removing pipe consists of equipment preparations, soil excavation, and backfill, which has significant meaning for waste disposal. The technologies used in this process is similar to technologies using in the construction process although the order of treatment is opposite to construction process. Based on this characteristic, emission calculation formulas of infrastructure construction are also suitable for pipe removal.

3. Materials recycling

As pipelines are made of great volumes of metals, the recycle of metals can bring economic and environmental benefits. We can simplify this process and assume that there are 40% of metal can be recycle, which means recyclable coefficient C_R equals to 40%.

Then, Eq. (4.18) is established to compute emission E_8 produced by recycle and disposal process, wherein E_1C_R in this formula is negative due to the recyclability of materials. The total emission produced by oil pipeline can be assessed through the computational results of E_1 to E_8.

$$E_8 = E_{\text{cleansing}} + E_3 + E_1C_R \tag{4.18}$$

4.3.5 Impact on Soil Environment

During the construction stage, the pipe trench will inevitably produce spoil. Soil that is originally in the space occupied by the pipe will be discarded. The volume of pipeline has a significant impact on the amount of excavated soil, and the amount of abandoned soil V_{as} is computed by Eq. (4.19).

$$V_{\text{as}} = \sum_{i=1}^{n} \pi(r_i + \delta_i + w_i)^2 L_i \tag{4.19}$$

Backfilling soil is inevitable to form a weak surface, which is another reason of causing soil erosion. It increases the chance to lead geological disasters containing landslide and collapse. In the practical engineering, 10–20 m is the normal size of working band of oil products pipeline, and Eq. (4.20) is proposed to compute soil erosion V_{sd}. There will be soil erosion modulus when constructing pipelines, and M_2 and M_1 represent the soil erosion modulus after and before the construction. T_i is the planned construction period, and F_i indicates the specific construction area [18, 19].

$$V_{\text{sd}} = \sum_{i=1}^{n} (M_2 - M_1)F_iT_i \tag{4.20}$$

Soil with outstanding carbon sequestration function is widely considered an ideal carbon container. Thus, soil abandonment and loss in process of pipeline excavation have potentially increased the amount of carbon storage in the soil. According to related studies [20], 20.5 Mg/ha is a suitable value of for the average soil organic carbon storage potential p_{loss}, then the amount of reduced carbon in this process can be calculated by Eq. (4.21).

$$E_{\text{loss}} = (V_{\text{as}} + V_{\text{sd}})p_{\text{loss}} \tag{4.21}$$

4.4 Life Cycle Analysis of Gas Pipelines

Differences in the life cycle of oil products and natural gas pipelines are mainly embodied in the operation stage. Thus, this section concentrates on the detailed calculation model of the operation stage. At this stage, three aspects of carbon emissions are considered: in the process of putting the pressure station into production, the damage in the process of equipment use and maintenance, and the leakage of pipeline damage due to accidents. Next, three parts are introduced in detail

4.4.1 Compressor Station

Compressor stations of the natural gas pipeline system provide most of the necessary energy for gas transmission, and it also produces a large number of carbon emissions. The energy consumption of compressors has a different calculation method than that of oil pumps. The hydraulic friction coefficient λ which impacts the energy consumption significantly of pipelines should be computed firstly, and the calculation formula can be expressed as follows:

$$\lambda = 0.009588 D_i^{-0.2} \tag{4.22}$$

Based on the hydraulic friction coefficient, the flow rate can be attained by Eq. (4.23), wherein Q is the volumetric flow rate and C_0 is the coefficient of pressure loss. P_Q and P_Z are the pressure of the start point and end point of pipelines. The compressibility of the gas in isothermal situation is represented by Z, and Δ indicates the relative density of natural gas. T is the gas temperature.

$$Q = C_0 \sqrt{\frac{(P_Q^2 - P_Z^2)d^5}{\lambda Z \Delta T L}} \tag{4.23}$$

Computing the energy consumption requires specific pressure ratio and flow rate. Here, the pressure distribution of pipelines are known. For the compressors, its energy consumption is also determined by motor speed n, and the mathematical formulas are expressed as follows:

$$Q = \frac{n}{n'} Q' \tag{4.24}$$

$$M = \frac{n}{n'} M' \tag{4.25}$$

$$N = \left(\frac{n}{n'}\right)^3 N' \tag{4.26}$$

N represents the power, and M is mass flow rate, while N', Q', n' refer the power, mass flow rate, and motor speed in work condition.

Based on Eqs. (4.24)–(4.26), we can attain Eq. (4.27); therefore, the power can be calculated by flow rate if the pressure ratio of compressor stations is known.

$$\frac{N}{N'} = \left(\frac{Q}{Q'}\right)^3 = \left(\frac{n}{n'}\right)^3 \tag{4.27}$$

Then, carbon emissions of compressor stations can be computed based on power consumption and conversion coefficient of carbon emission, as shown in Eq. (4.28). f_{op} refers the amount of carbon emissions caused by compressor stations, and t is the working time of compressors.

$$f_{\text{op}} = \sum_{i=1}^{k} N_i EF_{\text{elec}}\, t \tag{4.28}$$

4.4.2 Maintenance

The maintenance of equipment is the other main source of carbon emissions in operation stage that we cannot ignore, but it is hard to quantify it. Here, we adopt economic ratio method to calculate the carbon emission caused by maintenance. In the area of practical engineering economics, approximately 10% of the annual depreciation cost is equal to the maintenance cost. Based on this, the specific ratio between maintenance cost and total investment can be estimated. As economic cost is positive related to carbon emissions, we can assume that this ratio is also suitable for the carbon emissions. After completing the carbon emission calculation of all stages except maintenance process, the carbon emission of maintenance process can be calculated based the ratio. Specific compute formula is expressed as Eq. (4.29).

$$\frac{C_{\text{ma}}}{C_{\text{to}}} = \frac{f_{\text{ma}}}{f_{\text{to}}} \tag{4.29}$$

C_{ma} and f_{ma} are the economic cost and carbon emission of maintenance process, C_{to} refers the total investment of pipeline, and f_{to} is the carbon emission in total.

4.4.3 Leakage

Due to the great differences of physical properties between natural gas and oil, the carbon emission calculation method of leakage process for natural gas pipelines is also different when comparing to oil product pipelines. Two aspects including leakage probability and leakage amount are considered in the model of computing the carbon emission. In addition, we assume that the leakage gas all consists CH_4. According to statistics, the probability of leakage can be estimated. As shown in Table 4.1, the failure probability of natural gas pipelines is 0.157/1000 km per year.

The leakage of natural gas pipelines is affected by many factors, containing leakage speed and leakage duration. In general, the calculation model of leakage can be divided into three types: large hole leakage model, small hole leakage model, and pipeline fracture model. Statics shows that third party damage leads to 76% of pipeline leakage in China and 54% of natural gas pipeline leakage worldwide belongs to small hole leakage. Therefore, small hole leakage model is applied to compute the leakage volume of natural gas. The specific calculation method is expressed as follows:

When $P_2/P_a <$ CPR, the gas leakage rate m is:

$$M = C_d A P_2 \sqrt{\frac{W_g}{ZRT_2} \gamma \left(\frac{2}{\gamma+1}\right)^{\frac{\gamma+1}{\gamma-1}}} \tag{4.30}$$

When $P_2/P_a >$ CPR, the gas leakage rate m is:

$$M = C_d A P_2 \sqrt{\frac{2W_g}{ZRT_2} \frac{\gamma}{\gamma-1} \left[\left(\frac{P_a}{P_2}\right)^{\frac{2}{\gamma}} - \frac{P_a}{P_2}\right]^{\frac{\gamma+1}{\gamma}}} \tag{4.31}$$

CPR is the critical pressure ratio, A refers the leakage area, P_a is standard atmospheric pressure, C_d indicates the correction coefficient ranging in 0.61–1.0, the coefficient of polytropic is represented by γ, W_g is the molar mass of gas, and R refers the gas constant.

Table 4.1 Pipeline failure probability

Data sources	EGIG	CONCAWE	US DoT		CEPA
Research period	2004–2013	2001–2006	2002–2007		2004–2013
Type of transportation media	Gas pipeline	Oil pipeline	Oil pipeline	Gas pipeline	Gas/Oil pipeline
Pipeline failure probability (/1000 km·a)	0.157	0.280	0.340	0.089	0.140

Table 4.2 Leakage rate of pipeline with a different leakage hole diameter

Leakage hole diameter (m)	Leakage hole area (m^2)	Leakage (kg/s)
0.01	7.85×10^5	4.59
0.025	4.91×10^4	28.71
0.05	1.96×10^3	114.82
0.1	7.85×10^3	459.28

Based on above formulas, we computed the leakage amount of natural gas pipe with different leakage hole diameter, and computational results are shown in Table 4.2.

4.5 Life Cycle Analysis of Pipeline Samples

4.5.1 Oil Pipelines

Six pipelines are used as the objections of calculating the carbon emission, and Fig. 4.2 displays the routes of pipelines. Table 4.3 gives the basic parameters of these pipelines, including pipe length, diameter, and elevation. The physical properties of oil such as density and viscosity are shown in Table 4.4.

Base on the basic information of pipelines, the carbon emissions of pipelines can be computed by the LCA model, and Table 4.5 shows the computational results.

As Table 4.5 expressed, we use the CO_2e for delivering 10^4t products for 1 km to assess the emissions of each pipeline, which has more fairness. The results show that pipelines with larger distance will produce more carbon emissions. To comprehensively evaluate the impact of different links on carbon emission of different life stages, Table 4.6 displays the specific carbon emission of life stages. We can see that the construction of pipelines and pump stations is the main source of carbon emission. Meanwhile, the carbon emission of transmix treatment only takes a small part of total carbon emission. The infrastructure and intensity of regional power grid have a great impact on the difference between energy consumption and emission.

4.5.2 Gas Pipelines

Figure 4.3 shows the topology structure of the natural gas pipeline system, and the specific parameters including the length, diameter, and wall thickness of the pipe are given in Table 4.7.

The carbon emissions of natural gas pipelines can be calculated based on the above data, and the results are shown in Table 4.8.

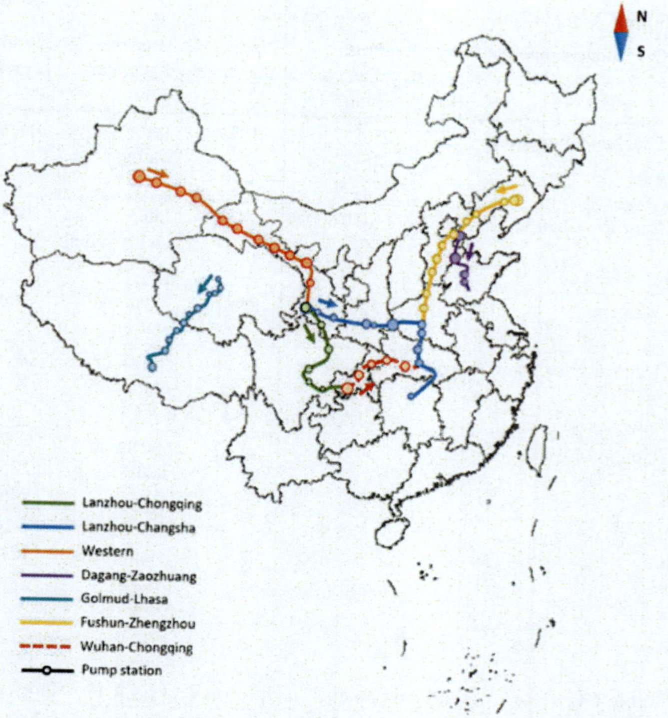

Fig. 4.2 Geographical diagram of oil products pipelines

The results show that 80% of carbon emission is produced by the manufacture stage, and the operation stage also has large amount of carbon emission, accounting for 1.1334×10^5t. However, the carbon emission of the other stages such as the construction stage and recycle stage is much smaller than manufacture stage and operation stage. These demonstrated that the main source of carbon emission is generated in pipeline construction of the first stage.

4.6 Conclusion

As climate issues gradually attract public attention worldwide, the operation and construction of pipeline networks have been attached with new energy-saving and emission-reduction targets. This chapter systematically sorts out and summarizes the carbon emission sources and environmental impacts of pipelines during the whole life cycle, and LCA models for oil products pipelines and natural gas pipelines are proposed in this chapter. For both oil products and natural gas pipeline systems, the results show that emissions generated by producing pipelines take the largest

Table 4.3 Input parameters of studied pipelines

Oil products pipeline	Pipe length (km)	Pipe diameter (mm)	Pipe elevation (m)	Transmix download station	Product types	Designed capacity (10^4 t/a)
Lanzhou-Chongqing	1247	508, 457.2, 323.9	−1329	Terminal station	0#, 92#	546
Lanzhou-Changsha	2087	610, 660, 508	−1471	Intermediate and terminal station	0#, 92#	570
Western	1858	559, 508	447	Terminal station	0#, 92#	440
Dagang-Zaozhuang	610	355.6, 273.1, 219.1	66	Terminal station	0#, 92#	285
Golmud-Lhasa	1119	323.9	588	Terminal station	−35#, −20#, 0#, 92#,95#	25
Fushun-Zhengzhou	1585	457, 508, 559, 610, 669	139	Intermediate station and Terminal station	−10#, 0#, 92#, 95#	800

Table 4.4 Viscosity of different products at different delivery temperatures

	Gasoline				Diesel	
Product number	−35#	−20#	−10#	0#	92#	95#
Density	0.823	0.834	0.841	0.847	0.724	0.725
Viscosity(20 °C)	6.016				0.613	

part of total carbon emissions, followed by the operation of the pumping station. It demonstrated that technologies which can reduce the carbon emission of steel manufacturing industry are the most potential solutions for reducing carbon emissions of pipeline system. Besides, optimal scheduling and operation of oil products and natural gas pipelines are also helpful for the reduction of carbon emissions.

Table 4.5 Emissions associated with each activity considered as part of the life cycle

	Lanzhou-Chongqing	Lanzhou-Changsha	Western	Dagang-Zaozhuang	Golmud-Lhasa	Fushun-Zhengzhou
Total CO_2e (t)	4.39×10^5	8.25×10^5	7.68×10^5	2.15×10^5	2.33×10^5	7.05×10^5
Unified CO_2e (tCO_2e/t km)	3.23×10^{-6}	3.47×10^{-6}	4.70×10^{-6}	6.19×10^{-6}	4.17×10^{-5}	2.78×10^{-5}

Table 4.6 Life cycle GHG emissions associated with different phases (t)

Oil Products Pipeline	Pipe manufacturing	Pipe construction	Pump station construction	Transmix processing	Pump operation	Fugitive emission	Recyclable content	Recovery operations
Lanzhou-Chongqing	6.51×10^5	136.89	1.25×10^3	26.11	1.62×10^5	1.48×10^4	3.91×10^5	853.60
Lanzhou-Changsha	1.49×10^6	303.37	3.16×10^3	81.21	2.09×10^5	1.54×10^4	8.92×10^5	1297.57
Western	1.15×10^6	227.10	2.30×10^3	54.58	2.90×10^5	1.19×10^4	6.93×10^5	3543.62
Dagang-Zaozhuang	1.52×10^5	85.79	6.01×10^3	20.88	1.43×10^5	7.72×10^3	9.14×10^4	3072.68
Golmud-Lhasa	4.55×10^5	140.38	0.90×10^3	7.04	4.60×10^4	677.46	2.73×10^5	3369.44
Fushun-Zhengzhou	1.14×10^6	218.35	3.06×10^3	68.05	2.21×10^5	2.17×10^4	6.84×10^4	2620.98

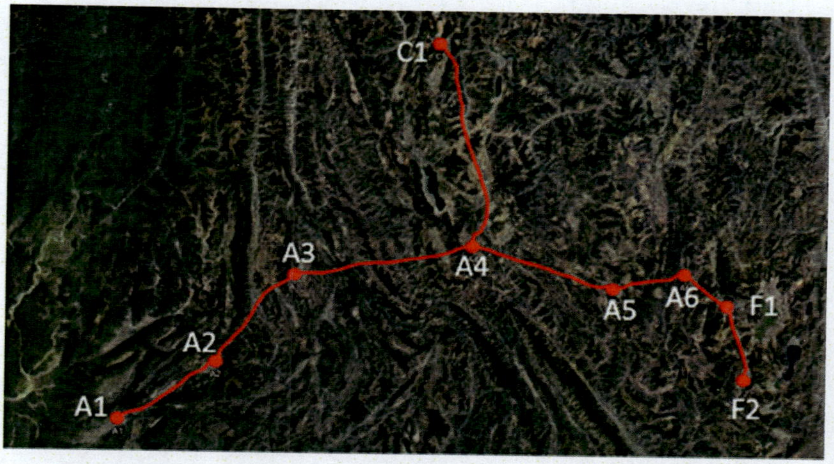

Fig. 4.3 Geographical diagram of natural gas pipelines in the case study

Table 4.7 Parameters of each case pipeline

Station	Download volume $10^8 m^3/a$	Pressure MPa	Type of station	Length/km	Quantity of steel used/t
A1	0	6.8	Pressurize	0	192,660
A2	0.22	/		104.61	
A3	0.89	7.9	Pressurize	140.25	
A4	0.09	/		153.12	
A5	1	/		154.64	
A6	0.1	6.0	Pressurize	47.82	
F1	11.68	/		42.78	10,355
F2	3.5	/		102.86	
C1	1.57	/		200.84	6081.93

Table 4.8 Carbon dioxide emissions in each stage of each case

Stage		Carbon emission/t	Total carbon emissions in each stage /t
Manufacture stage	Pipeline production	6.9838×10^5	7.3163×10^5
	Equipment production	770.00	
	Transportation process	3.2472×10^4	
Construction stage	Excavation and backfilling of the pipe trench	481.42	2296.19
	Pipeline construction	14.77	
	Station construction	1800.00	
Operation stage	Operation of pressure station	4.0179×10^4	1.1334×10^5
	Leakage (methane)	/	
	Maintain	7.3162×10^4	
Recycle stage	Pipeline removal	496.19	2.6474×10^4
	Transportation process	2.5978×10^4	

References

1. Wang K, Wang C, Lu X, Chen J. Scenario analysis on CO2 emissions reduction potential in China's iron and steel industry. Energy Policy. 2007;35:2320–35. https://doi.org/10.1016/j.enpol.2006.08.007.
2. Ren Z. Research of building energy system carbon emissions accounting base on LCA: Dalian University of Technology; 2014.
3. Samsudin MSF, Dell'Olio M, Leong KH, Ahamid Z, Varley RJ. Adhesives performance of 3-layer PE pipe coatings: Effects of MAH loading, PE particles size, coating interval time and service temperature. Progress in Organic Coatings. 2016;99:157–65. https://doi.org/10.1016/j.porgcoat.2016.05.013..
4. NANJO, Takao, IMANISHI, Etsujiro, OOTANI, Kazuhiro, et al. Simulation and Evaluation Technique for Power System and Related Energy Saving on Hydraulic Excavator. Kobelco Technology Review. 2007:28–34. https://www.kobelco.co.jp/english/ktr/pdf/ktr_27/028-034.pdf.
5. Liu M, Qian F, Zhan X. Calculation Model for Energy Carbon Emission of Building Material Transportation. 2010 International Conference on E-Product E-Service and E-Entertainment2010. p. 1–3. https://doi.org/10.1109/ICEEE.2010.5660454.
6. Jocanovic M, Agarski B, Karanovic V, Orosnjak M, Stankovski S. LCA/LCC Model for Evaluation of Pump Units in Water Distribution Systems. Symmetry. 2019;11:1181-. https://doi.org/10.3390/sym11091181.
7. Nault J, Papa F. Lifecycle Assessment of a Water Distribution System Pump. Journal of Water Resources Planning & Management. 2015;141:A4015004. https://doi.org/10.1061/(ASCE)WR.1943-5452.0000546.
8. Liao Q, Castro PM, Liang Y, Zhang H. Computationally Efficient MILP Model for Scheduling a Branched Multiproduct Pipeline System. Industrial & Engineering Chemistry Research. 2019b. https://doi.org/10.1021/acs.iecr.8b06490.
9. Liao Q, Castro PM, Liang Y, Zhang H. New batch-centric model for detailed scheduling and inventory management of mesh pipeline networks. Computers & Chemical Engineering. 2019c;130:106568. https://doi.org/10.1016/j.compchemeng.2019.106568.

10. Austin JE, Palfrey JR. Mixing of Miscible but Dissimilar Liquids in Serial Flow in a Pipeline. ARCHIVE Proceedings of the Institution of Mechanical Engineers 1847–1982 (vols 1–196). 1963;178:377–95. https://doi.org/10.1243/PIME_PROC_1963_178_030_02.
11. Rachid FBF, De Araujo JHC, Baptista RMJJoFE. Predicting Mixing Volumes in Serial Transport in Pipelines. 2002;124:528–34. https://doi.org/10.1115/1.1459078.
12. Melo S, Rachid F. Computing Transmix in Complex Batch Transfers via the Mixing-Volume-Equivalent-Pipe Concept2010. https://doi.org/10.1115/IPC2010-31607.
13. Milazzo MF, Vianello C, Maschio G. Uncertainties in QRA: Analysis of losses of containment from piping and implications on risk prevention and mitigation. Journal of Loss Prevention in the Process Industries. 2015;36:98–107. https://doi.org/10.1016/j.jlp.2015.05.016.
14. Eggleston S, Buendia, L., Miwa, K., Ngara, T., Tanabe, K. 2006 IPCC Guidelines for National Greenhouse Gas Inventories. Institute for Global Environmental Strategies Hayama. Institute for Global Environmental Strategies Hayama, Japan. 2006.
15. IPCC. Fourth Assessment Report (AR4): Climate Change 2007: Synthesis Report. 2007.
16. Di Lullo G, Oni AO, Gemechu E, Kumar A. Developing a greenhouse gas life cycle assessment framework for natural gas transmission pipelines. Journal of Natural Gas Science and Engineering. 2020;75:103136. https://doi.org/10.1016/j.jngse.2019.103136.
17. Jessica M.Swanson TK, Pete Poohkay. Environmental considerations for pipeline abandonment. pipes and pipelines international. 2011:p. 1–6. https://doi.org/10.1115/IPC2010-31669.
18. Guo Q, Hao Y, Liu B. Rates of soil erosion in China: A study based on runoff plot data. CATENA. 2015;124:68–76. https://doi.org/10.1016/j.catena.2014.08.013.
19. Zhou ZC, Shangguan ZP, Zhao D. Modeling vegetation coverage and soil erosion in the Loess Plateau Area of China. Ecological Modelling. 2006;198:263–8. https://doi.org/10.1016/j.ecolmodel.2006.04.019.
20. Kayet N, Pathak K, Chakrabarty A, Sahoo S. Evaluation of soil loss estimation using the RUSLE model and SCS-CN method in hillslope mining areas. International Soil and Water Conservation Research. 2018;6:31–42. https://doi.org/10.1016/j.iswcr.2017.11.002.

Chapter 5
Operation Condition Monitoring for Pipeline

Li Zhang

Abstract Monitoring pipeline operating conditions is a vital component in pipeline safety and risk management. Although the SCADA system stores a large amount of operational data, the data lacks associated condition labels, making it difficult to mine. Furthermore, the operating circumstances of the multi-product pipeline vary often, and identification and monitoring by on-site employees are prone to error, so the pipeline's operating conditions cannot be reliably identified. To address the aforementioned challenges, this chapter presents semi-supervised learning for operating condition identification. The findings show that semi-supervised learning has more stability and improved performance regardless of how the neural network is built. The suggested technique may be utilized as a decision-making tool for monitoring and identifying multi-product pipeline operating conditions.

5.1 Introduction

Oil and gas transmission is only one of the many jobs taken on by the pipeline industry. For reasons of both national economic and military importance, the reliable functioning of pipeline transit is crucial. Consequently, identifying the operating state of long-distance oil (gas) pipelines is an effective method of averting potentially catastrophic mishaps during oil (gas) transfer. However, due to the many transportation options and the nature of everyday activities, operating status is always shifting. Some operating states that may pose safety risks may only be determined by pressure curves, with the help of human inspections at the station, which presents a difficulty in the event of an incomplete data label. It's a slow and time-sensitive method of evaluation. It is also challenging to create an accurate causal network connecting neighboring stations in a vast and complicated long-distance pipeline network. Meanwhile, it's difficult to depend just on expert expertise when trying to establish the prior probability between various causal connections.

L. Zhang (✉)
National Engineering Laboratory for Pipeline Safety; Beijing Key Laboratory of Urban Oil and Gas Distribution Technology, China University of Petroleum, 102249 Beijing, China
e-mail: Zhanglili1229@hotmail.com

H. Su et al. (eds.), *Advanced Intelligent Pipeline Management Technology*,
https://doi.org/10.1007/978-981-19-9899-7_5

When scientists first began their investigations, they focused mostly on finding ways to detect breaches in pipelines. Osegi [1] suggested a pipeline data monitoring system based on a very sparse predictive deviation learning algorithm to address the shortcomings of conventional physical–mathematical models, namely their lack of timeliness and predictive ability (p-DLA). Our approach is distinct because it makes use of actual data from pipeline operations to do its modeling and validation, rather than hypothetical data. Signal processing algorithms are used by Kim et al. [2]. to pinpoint the exact site of pipeline leaks. However, the sensor data was impacted by noise, and the sensor network could not cover the whole pipeline network, leading to a significant false alert rate. The dynamic pressure waveform was analyzed by Liu et al. [3]. using wavelet analysis and empirical mode decomposition. To locate the pipeline leak signal, the support vector machine technique was used for. While the aforementioned techniques all identify an unusual operating situation, none of them can generalize to other cases. While some of our work does fall under the umbrella of anomaly detection, we put special emphasis on identifying change points that are better characterized by operational states; we also work on improving methods for segmenting data from operation records; and finally, we work to identify operational patterns across time.

Fundamental supervised learning criteria cannot be satisfied in oil and gas pipeline systems due to a lack of trustworthy fault data and event annotation over several time periods. Consequently, there has not been significant amount of research into categorizing and identifying gas and oil pipeline operating states. The BN based on the complete pipeline system must be built if it is to be used in studies of recognizing the operational state of pipelines. The building process is challenging because it requires specialized knowledge and skill. In addition, incorporating expert opinion seldom yields accurate prior probability of occurrences. In conclusion, it is challenging to apply the BN to infer and recognize the operating modes of pipelines network systems based on the existing state of accessible pipeline network data and event information.

In this chapter, we present a data-driven state identification model that uses data mining techniques to identify the operational condition of petroleum and natural gas pipeline systems from two different vantage points: the equipment level and the subtransmission station level. It's possible that additional huge and complicated systems might benefit from using the graph structure feature-based approach.

To identify out-of-the-ordinary pressure changes in pipeline systems during operation, the change point detection technique [4] is used. In addition, the approach may derive the partitioning of time series data using transition points as boundaries, which aids in reacting to the consequences of missing certain actual occurrences because of faulty sensors or incorrect settings. Finally, the operating modes of the pipeline are recognized using statistical feature-based, structure feature-based, and interval-based time series classification approaches. We compare and explain the categorization algorithms based on four distinct operational situations to demonstrate the viability of the methodological framework.

5.2 Methodology

This machine learning-based approach to state recognition has two goals. First, it seeks to identify transitions in pipeline operation through unsupervised learning techniques. Second, oil pipeline operational conditions can be determined by data-driven analysis. Figure 5.1 illustrates the connection between the two problems. Each of the aforementioned processes is broken out into further depth below.

5.2.1 *Method for Detecting Changes in Operational States*

This section introduces the cumulative sum approach, which may be used to identify transitions between modes in a subtime series. It is possible to segment the time series at the observed transition points. We also provide metrics for assessing the quality of the tests. We also compared the change point detection approach with a control group comprised of isolation forests to further understand its benefits in this specific case study.

5.2.1.1 Change Point Detection Method

A state change identification model can be established to clarify the various operational changes experienced by the pipeline and even detect possible abnormal

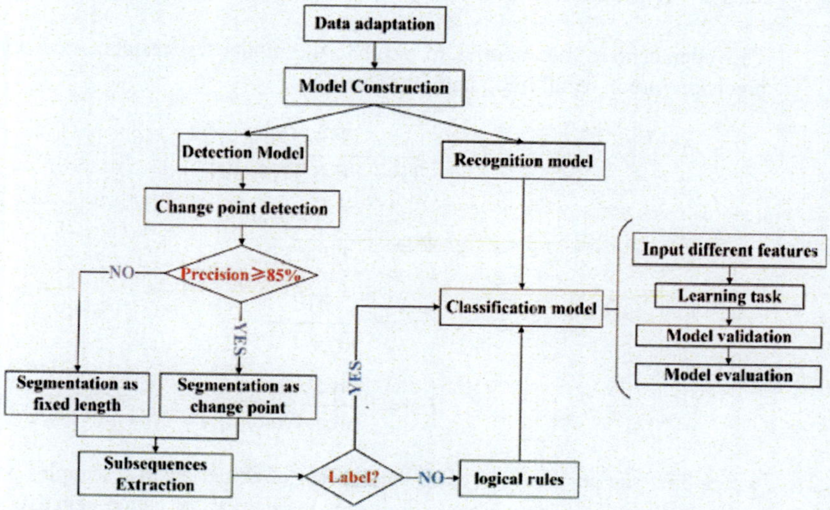

Fig. 5.1 Flowchart of methodology

state changes during pipeline operation. Combined with the event records from the SCADA system, the events that lead to changes in the pipeline's operating state can be quickly determined over different time periods.

When the pipeline operating state changes, the probability distribution of the pressure time series changes abruptly in two consecutive intervals, and since the two intervals are moving, a state change point is considered to exist when the probability distributions of the two intervals are significantly different. The method to achieve the above determination process is shown in Eq. 5.1, and this method is known as the cumulative sum (CUSUM) test method [5].

$$\begin{cases} s_t = x_t - x_{t-1} \\ g_t^+ = \max\left(g_{t-1}^+ + s_t - v, 0\right) \\ g_t^- = \max\left(g_{t-1}^+ - s_t - v, 0\right) \end{cases} \tag{5.1}$$

$$\text{if } g_t^+ > h \text{or} g_t^- > h : \begin{cases} t_{\text{alarm}} = t \\ g_t^+ = 0 \\ g_t^- = 0 \end{cases}$$

where g_t^+ and g_t^- denote the cumulative sum of positive and negative changes, respectively. h and v denote user-defined thresholds and correction parameters, respectively, whose purpose is to weigh the number of true and false alarms. Equation 5.1 indicates that if g_t^+ and g_t^- exceed the user-defined threshold (h), an alarm is raised and the cumulative sum is then reset and the next detection is made [6].

5.2.1.2 Evaluation Indicators

This chapter uses four metrics to evaluate the model test results: accuracy (acc), precision (prec), recall (rec), and $F1$ value [7].

$$\text{acc} = \frac{\text{TP} + \text{TN}}{\text{TP} + \text{FP} + \text{TN} + \text{FN}} \tag{5.2}$$

$$\text{prec} = \frac{\text{TP}}{\text{TP} + \text{FP}} \tag{5.3}$$

$$\text{rec} = \frac{\text{TP}}{\text{TP} + \text{FN}} \tag{5.4}$$

$$F1 = 2\frac{\text{prec} \times \text{rec}}{\text{prec} + \text{rec}} \tag{5.5}$$

where TP (true positive) indicates the number of actual positive samples predicted to be positive; TN (true negative) indicates the number of actual negative samples predicted to be negative; FP (false negative) indicates the number of actual positive samples predicted to be negative; and FN (false positive) indicates the number of actual negative samples predicted to be positive.

5.2.2 Characteristic Representation

A feature vector of low dimension is used in feature-based classification and recognition algorithms to uncover the hidden information in a subsequence. In this part, we'll go over two different ways of representing features: time–frequency domain statistical features and complex network structure features. In order to characterize the complexity, volatility, and smoothness of the input subtime series, statistical features perform mathematical operations on the data. The statistical features used include the mean, kurtosis, skewness, standard deviation value, variance, complexity, approximate entropy, and absolute energy. The complex network structure features are obtained by converting the subtime series into the form of a complex network graph; the specific conversion and feature extraction process are shown in Fig. 5.2.

A pair of nodes is considered to have an edge if and only if it is possible to draw a straight line between them, regardless of whether or not any additional data is between them, according to the conversion rule [8]. The process of visibility graph conversion is shown in the left corner of Fig. 5.2. Eq. 5.6 also provides a definition for the laws of transformation.

$$y_c < y_b + (y_a - y_b)(t_b - t_c/t_b - t_a) \tag{5.6}$$

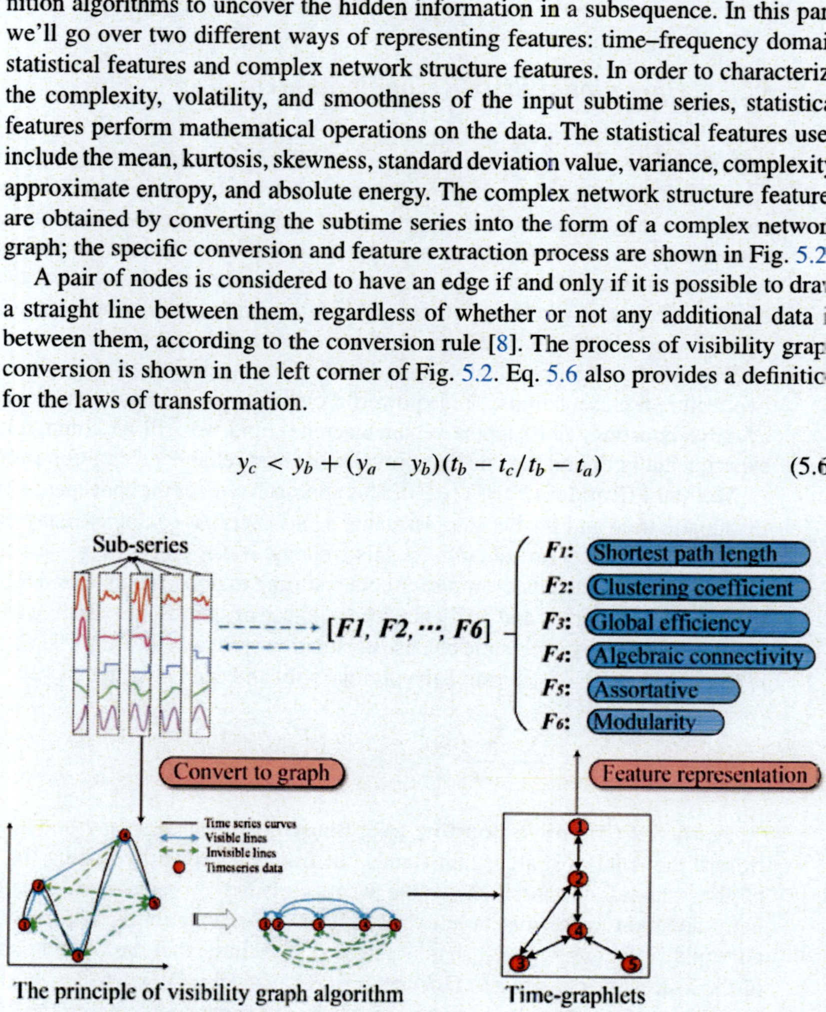

Fig. 5.2 Process of characterizing the structure of complex networks

where t_a, t_b, and t_c represent different times in time series; y_a, y_b, and y_c mean the different values corresponding to the different times.

The advantage of using complex network structure features for characterization is that this allows the reduced dimensional features to not only feature the shape of the time series, but also contain the physical information behind the data to be conveyed. In this study, six graph network features, shortest path length, clustering efficiency, global efficiency, algebraic connectivity, homomorphism, and modularity, are used for characterization.

5.2.3 Operational State Recognition Method

In this investigation, we extract features for differential pressure by integrating the operational characteristics of the pipeline with special attention to rapid changes and volatility. Using features of the statistical and graph structures, we may identify the operational states. The objective is to utilize classification techniques to determine what state a subsequence belongs to by embedding it into a set of characteristics. As a benchmark, we also tried both the interval-based and dictionary-based categorization strategies to help us choose the most effective recognition approach.

We begin by selecting a feature-based method for categorizing the outcomes of the two feature representations. Three powerful classifiers were chosen (random forest, adaptive boosting, and support vector machine), and we will be comparing them experimentally to find the one that provides the best reliable recognition results.

Method 1 (Random Forest (RF)). This approach utilizes the concept of ensemble learning to train and predict samples using a classifier that combines many decision trees. Large datasets are suitable for this strategy. It can process high-dimensional characteristics in the input sample without resorting to dimensionality reduction. On top of that, its accuracy and resilience surpass those of standalone classifiers, making it not only generally accessible but also useful in a variety of contexts. The Gini index is often used in RF to determine the optimal split and can be written as [9]:

$$\sum\sum_{j\neq i}(f(C_i,T)/|T|)\left(f\left(C_j,T\right)/|T|\right) \tag{5.7}$$

Method 2 (Adaptive Boosting (AdaBoost)). Among the many boosting algorithms, the AdaBoost algorithm stands out from the crowd by raising the weight of the samples misclassified by the weak classifier in each round, so drawing more attention to the misclassified data. The implementation is as follows: given a sample $T = \{(x_1, y_1), (x_2, y_2), \dots, (x_m, y_m)\}$, assume that the weight coefficient of the k-th weak classifier is $D(k) = (w_{k1}, w_{k2}, \dots, w_{km})$, where $w_{ki} = 1/m$ for $i = 1, 2, \dots, m$, , then the corresponding sample weight coefficient of the $(k + 1)$th weak classifier is defined as [5]:

$$w_{k+1,i} = \frac{w_{ki}}{Z_K}\exp(-\alpha_k y_i G_k(x_i)) \tag{5.8}$$

$$Z_K = \sum_{i}^{m} w_{ki}\exp(-\alpha_k y_i G_k(x_i)) \tag{5.9}$$

where Z_K is the normalization factor, α_k is the weight coefficient of the k th weak classifier $G_k(x)$.

Furthermore, AdaBoost also uses a weighted majority voting system. Votes are given more weight by weak classifiers that have low classification error rates. An exhaustive set of robust classifiers is given by Freund and Schapire [5]:

$$f(x) = \mathrm{sign}\left(\sum_{k=1}^{K}\alpha_k G_k(x)\right) \tag{5.10}$$

Method 3 (Support vector machine (SVM)). As a high-accuracy classifier, the SVM approach has found several uses in the area of pattern recognition. The primary principle behind SVM is to categorize areas by locating isolated hyperplanes that may split the greatest geometric interval of the training datasets [10]. For a given dataset $T = \{(x_1, y_1), (x_2, y_2), \ldots, (x_n, y_n)\}$, where $y_i \in \{+1, -1\}$, is binary class label, $i = 1, 2, \ldots, n$. If the two classes of the set may be split by a specific spatial hyperplane, then the ideal hyperplane maximizes the margins. This is true even if the previous distribution of the data is unknown. The definition of the categorization decision function is as follows [10]:

$$f(x) = \mathrm{sign}\left(\sum_{i=1}^{N}\alpha_i^* y_i K(x, x_i) + b^*\right) \tag{5.11}$$

where the optimal values for α^* and b^* can be found by solving the following quadratic programming problem [11]:

$$\min_{\alpha}\frac{1}{2}\sum_{i=1}^{N}\sum_{j=1}^{N}\alpha_i\alpha_j y_i y_j K(x_i, x_j) - \sum_{i=1}^{N}\alpha_i \tag{5.12}$$

$$\sum_{(i=1)}^{N}\alpha y_i = 0, (0 \leq_i \alpha_i\, C, i = 1, 2, \ldots, N) \tag{5.13}$$

$$b^* = y_j - \sum_{i=1}^{N}\alpha_i^* y_i K(\boldsymbol{x}_i, \boldsymbol{x}_j) \tag{5.14}$$

where $K(x, z)$ represents the kernel function and C is the penalty parameter.

5.3 Case Study

5.3.1 *Operational State Detection*

To prove the viability of the change point detection approach based on the product pipeline transportation system, we set up CUSUM-based detection studies for various situations. Methods of application and scenario-specific parameter analyses are outlined below.

Valve status detection is one of an important scenarios in the system. The differential in pressure between the input and output of the valve is the focus of this investigation. The whole opening and closing of the valve is chosen as the criteria for picking the historical event description since the change of valve state is a continuous process, and this experiment only focuses on the first instant when the valve state is altered. To compensate for the lag in sensor data during the state transition process, a maximum acceptable time error of 60 s was established. Each valve opening and closing action has its own distinct set of change points since the period between neighboring change points was set to 3600 s.

Figure 5.3 displays the outcomes of optimizing the detection technique for its two primary parameters, h and v. The accuracy and recall are unaffected by the h adjustments. That's because all possible values of truth are included in the detection findings. When comparing the accuracy at various values, it becomes clear that $h = 0.04$ and $v = 5 \times 10^{-7}$ result in the fewest false negatives. However, the $F1$ value is more than 85.7%. Hence, the parameters combination is the best parameter configuration for case 1.

The results of the evaluation of change point detection are shown in Table 5.1. As can be seen from Table 5.1, the change point identification method has a lower miss rate and better overall performance (as can also be seen from the $F1$ values). Compared to the isolated forest method, the miss rate is reduced by 26% in the valve condition detection application scenario. On the other hand, in the absence of data labelling, the results obtained by the change point identification method can be used as a basis for state segmentation of time series data. It can provide more accurate identification of large and complex systems that lack data identification or provide a basis for early warning of different modes of operation.

5.3.2 *Recognition of Pipeline Operational State on Single Station*

Classification studies include five states, including valve internal leakage (Scenario S1), valve switch (Scenario S2), pigging (Scenario S3), and pump shutdown (Scenario S4). The experimental findings demonstrate that several categorization models are relevant to various pipeline operating conditions in the process of

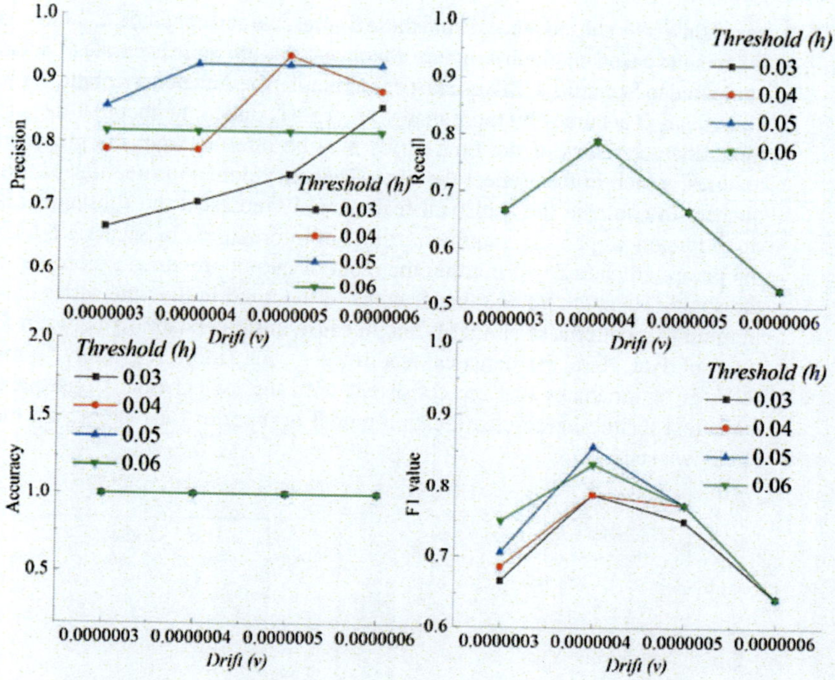

Fig. 5.3 Parameter optimization results for valve status detection

Table 5.1 Evaluation results of two method in valve status detection and pump shutdown

Scenario	Evaluation metrics	Change point detection	Isolation forest detection
valve status detection	Precision	0.9375	0.9
	Recall	0.7895	0.5294
	Accuracy	0.999	0.999
	$F1$	0.8571	0.6667
pump shutdown	Precision	0.912	0.909
	Recall	0.554	0.383
	Accuracy	0.999	0.999
	$F1$	0.689	0.539

addressing actual engineering challenges. Classification based on graph structural characteristics has better definition accuracy and more generalizability.

In Figs. 5.4a, 5.5a, 5.6a and 5.7a, we have a side-by-side comparison of the two feature-based categorization models. The model determines the state of operations based on a 60-s subsequence. From these figures, we can conclude that the classification results based on graph structure characteristics are more accurate in Scenario S1 compared to Scenario S3. To use just one example, the AdaBoost technique improved classification accuracy by more than 90%. This is due to the fact that the shape and trend characteristics of the time series may be inferred from the graph structure features, which in turn reflect the time–frequency domain properties of the subsequence. In contrast, the statistical features only represent the subsequence's local time-domain properties, resulting in a loss of information. In Scenario S4, however, the pump will shut down if either the input or output pressures exceed the set alert threshold. There are no defining features of the trend in the time series data. Keep in mind that patterns of similarity might cause misunderstanding between different types of data. Here, the statistical feature vector may more accurately represent the features of the states that are consistent with the logical rule. Consequently, the statistical feature-based classification model is superior for recognizing the pump shutdown state.

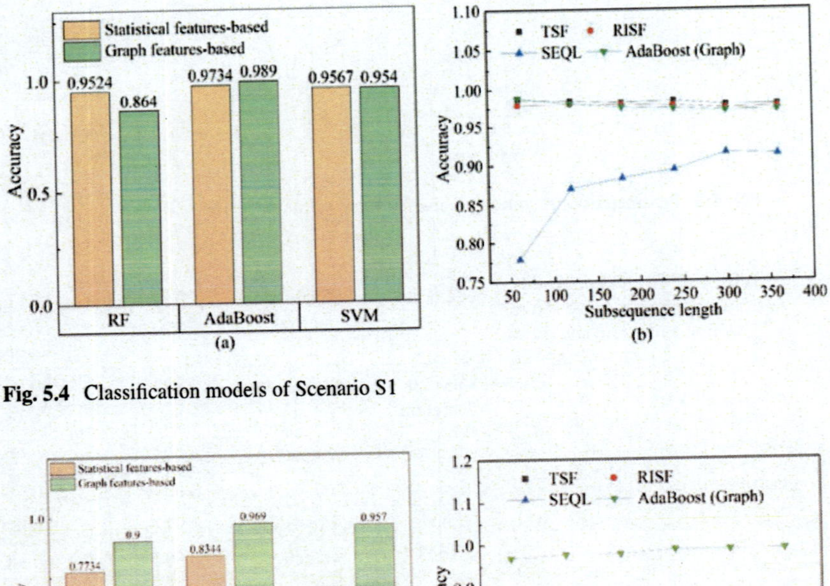

Fig. 5.4 Classification models of Scenario S1

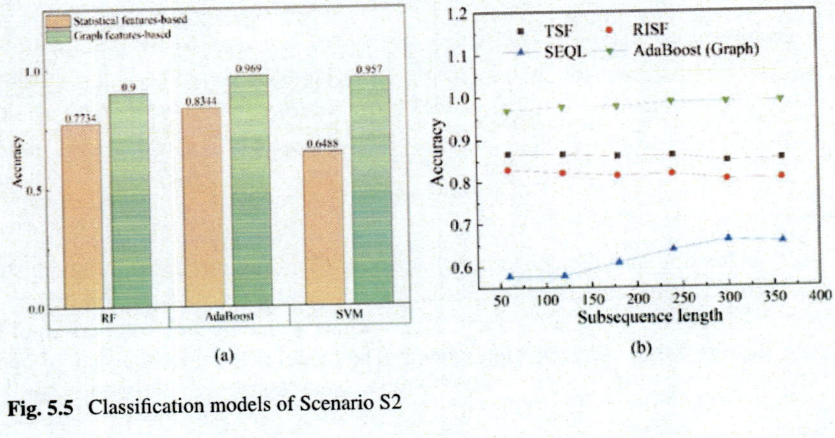

Fig. 5.5 Classification models of Scenario S2

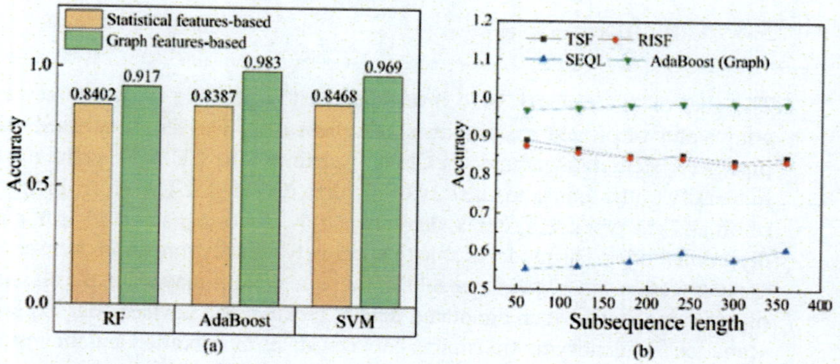

Fig. 5.6 Classification models of Scenario S3

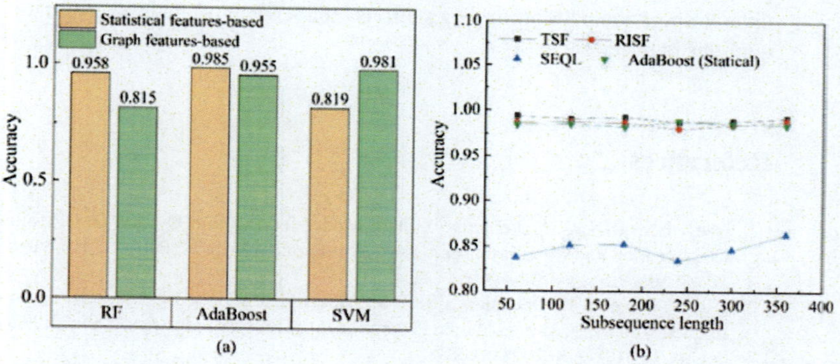

Fig. 5.7 Classification models of Scenario S4

Figures 5.4b, 5.5b, 5.6b and 5.7b contrast the AdaBoost model's and the time series-based classification model's classification accuracy. The influence of varying subsequence lengths on the final classification results is also shown in these figures. The figures show that the SEQL model is affected by the length of the input subsequence and has the worst classification accuracy. While the model does make use of SAX symbolic representation, it suffers from relatively poor classification accuracy since smaller data fluctuations are obscured by the symbolic interval, leading to lost information. However, the accuracy with a wide range of symbol intervals improves and the number of symbols grows with a longer sequence. The TSF and RISF classification accuracies are comparable. Our four examples show that their accuracy is more than 80% regardless of the length of the subsequence. In all four cases, the integrated AdaBoost classification model using graph features achieved the maximum accuracy while being mostly immune to variations in input sequence length.

5.4 Conclusion

The goal of this research is to investigate if it is possible to determine the operating status of oil and gas pipelines using just data. The report is based on actual pipeline system data gathered in China over time. The CUSUM technique is used to identify transitions in the history of pipeline operations. Therefore, in the context of oil and gas pipeline systems, demonstrate the abnormal status for our suggested detection model. The model is able to accurately identify transitions between equipment operating states, and the results of these detections may aid in the classification of operating states when annotated data is lacking. We devised four experimental scenarios for identifying transitions between states of operation and the results have been compared of those using a feature set, an interval set, and a time series dictionary. The classification approach based on graph structure characteristics is the most generalizable and accurate. Last but not least, the classification model can recognize diverse operation patterns among stations independent of the impact and topological qualities between them.

References

1. Osegi EN. p-DLA: A Predictive System Model for Onshore Oil and Gas Pipeline Dataset Classification and Monitoring-Part 1. arXiv preprint arXiv:170100040. 2016. https://doi.org/10.48550/arXiv.1701.00040.
2. Kim M-S, Lee S-K. Detection of leak acoustic signal in buried gas pipe based on the time–frequency analysis. Journal of Loss Prevention in the Process Industries. 2009;22:990–4. https://doi.org/10.1016/j.jlp.2008.08.009.
3. Liu W, Liu H. Research on Automatic Identification for the Leakage Signal of Petroleum Pipeline. Sensors & Transducers. 2013;21:147.
4. Page ES. Continuous inspection schemes. Biometrika. 1954;41:100–15. https://doi.org/10.2307/2333009.
5. Freund Y, Schapire RE. A decision-theoretic generalization of on-line learning and an application to boosting. Journal of computer and system sciences. 1997;55:119–39. https://doi.org/10.1006/jcss.1997.1504.
6. Dietterich TG. An experimental comparison of three methods for constructing ensembles of decision trees: Bagging, boosting, and randomization. Machine learning. 2000;40:139–57. https://doi.org/10.1023/A:1007607513941.
7. Liu L, Peng Y, Wang S, Liu M, Huang Z. Complex activity recognition using time series pattern dictionary learned from ubiquitous sensors. Information Sciences. 2016;340:41–57. https://doi.org/10.1016/j.ins.2016.01.020.
8. Lacasa L, Luque B, Ballesteros F, Luque J, Nuño Juan C. From time series to complex networks: The visibility graph. Proceedings of the National Academy of Sciences. 2008;105:4972–5. https://doi.org/10.1073/pnas.0709247105.
9. Pal M. Random forest classifier for remote sensing classification. International journal of remote sensing. 2005;26:217–22. https://doi.org/10.1080/01431160412331269698.

10. Schuldt C, Laptev I, Caputo B. (2004). *Recognizing human actions: a local SVM approach*. Proceedings of the 17th International Conference on Pattern Recognition, 2004 ICPR 2004. doi: https://doi.org/10.1109/ICPR.2004.1334462.
11. Rocco S CM, Zio E. A support vector machine integrated system for the classification of operation anomalies in nuclear components and systems. Reliability Engineering & System Safety. 2007;92:593–600. https://doi.org/10.1016/j.ress.2006.02.003.

Chapter 6
Operation Condition Prediction for Pipeline

Li Zhang and Huai Su

Abstract The present framework for managing long-distance oil and gas transmission pipelines is encountering difficulties as a result of the elements that are becoming more complicated, unpredictable, and time-dependent. It requires thorough system characteristics information, as well as precise beginning and boundary conditions, in order to be effective. In this chapter, we suggest using the deep learning approach to the operation and administration of the natural gas transmission system in an effort to get around these issues. The suggested approach enables efficient and reliable forecasts, particularly under unusual circumstances. The findings demonstrate that the suggested technique can operate the gas pipeline system effectively and efficiently while also making precise real-time forecasts beneficial for decreasing future operational losses.

6.1 Introduction

From sources through pipeline networks to the various clients, natural gas travels a great distance. Operations that are accurate and timely guarantee consistent client supply. This calls for exact timing knowledge of the system's and its parts' circumstances [1]. Therefore, the use of effective methods to accurately predict the dynamic response of a system and the future state of its components is a must.

From a broader perspective, we can see how advancements in AI, ML, and big data are altering how the energy sector is envisioned. Various studies have been conducted in an effort to address the many problems that exist in today's energy systems, such as system dependability and stability [2], operation efficiency and cost control [3], renewable energy management [4], and environmental concerns [5]. Various energy systems have benefited from research into the use of forecasting,

L. Zhang · H. Su (✉)
National Engineering Laboratory for Pipeline Safety; Beijing Key Laboratory of Urban Oil and Gas Distribution Technology, China University of Petroleum, 102249 Beijing, China
e-mail: suhuai@cup.edu.cn

L. Zhang
e-mail: Zhanglili1229@hotmail.com

H. Su et al. (eds.), *Advanced Intelligent Pipeline Management Technology*,
https://doi.org/10.1007/978-981-19-9899-7_6

categorization, and optimization approaches. These improvements have been felt by regulators, consumers, and operators alike. In addition, recurrent and cascade neural networks are competitive alternatives for simulating the future of dynamic systems [6].

Supervised control and data acquisition (SCADA) systems create and collect massive volumes of data on the operation, device condition, and gas consumption of natural gas pipeline network systems. There have been various attempts to estimate and anticipate gas demand in real time, online, using this huge data. In order to evaluate the live status of various devices for pipelines, most online state estimators are built on top of filter models [7]. Machine learning has attracted a great deal of scholarly interest in estimating gas demand, both in terms of scientific research and practical applications [8]. Foreseeing future needs in natural gas use has inspired the creation of several algorithms.

Nowadays, deep learning methods have received widespread attention from companies and academics. The deep structure is used to extract features hierarchically, allowing the dynamic and inherent properties of a particular pipeline system to be identified from the data [9]. It has been demonstrated that the complex dynamics of gas pipeline networks can be learnt autonomously by neural network black-box models without a priori information to appropriately predict the real-time operating state of the system. Although various black box models based on neural networks have been used for modeling and analysis of oil and gas pipelines, most of these studies have focused on single equipment such as compressors or on the analysis of pipeline safety [10]. Research on the use of deep learning methods to analyze the dynamics of complex pipe networks is yet to be further investigated.

An encoder-based deep learning approach is developed in this chapter to predict real-time changes and equipment status of complex gas transmission networks. To acquire knowledge of the dynamic aspects of the system, the stacked autoencoder model is trained layer by layer using a greedy algorithm, which is then used for prediction at the end of training. Consideration of temporal and geographical correlations is included into the model. Deep learning is combined with theories of network structure controllability and data windows to maximize data use.

6.2 Methodology

6.2.1 *Autoencoder*

Autoencoders differ from traditional neural networks in that they can reconstruct their inputs at the output layer. Figure 6.1 displays a sample autoencoder model that consists of an input layer, an output layer, and a hidden layer. Taken $\boldsymbol{x} = \{x_1, x_2, \ldots, x_n\}$ as an example, the process of encoded and decoded is defined as below:

$$\mathbf{y}(\mathbf{x}) = f(\mathbf{W}_1\mathbf{x} + \mathbf{b}) \tag{6.1}$$

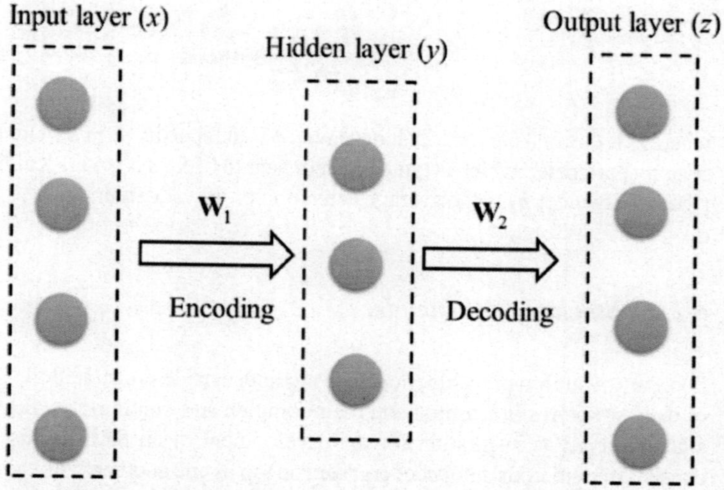

Fig. 6.1 Construction of the autoencoder sample model

$$\mathbf{z}(x) = g(\mathbf{W}_2\mathbf{y}(x) + \mathbf{c}) \tag{6.2}$$

$$f(x) = g(x)\frac{1}{1 + e^{-x}} \tag{6.3}$$

where $\mathbf{W}_1$ and $\mathbf{W}_2$ represent the weighting matrix and the decoding matrix respectively; **b** and **c** represent the encoding bias and decoding bias, respectively.

It is possible to determine the model parameters θ by reducing the reconstruction error (E),

$$\theta = \arg_\theta \min E(\mathbf{x}, \mathbf{z}) = \arg_\theta \min \frac{1}{2}\sum_{i=1}^{n} \|x_i - z(x_i)\|^2 \tag{6.4}$$

The above method can be used to learn the recognition function if the size of the input layer is smaller than the size of the hidden layer, so that sparsity restrictions are added to the objective function. But this also leads to a shift from autoencoders to alternate autoencoders. Therefore, the reconstruction error minimization problem can be calculated as follows:

$$S = E(\mathbf{x}, \mathbf{z}) + \phi \sum_{j=1}^{H} KL(\rho \| \hat{\rho}_j) \tag{6.5}$$

$$KL(\rho \| \hat{\rho}_j) = \rho \log \frac{\rho}{\hat{\rho}_j} + (1 - \rho) \log \frac{1 - \rho}{1 - \hat{\rho}_j} \tag{6.6}$$

$$\hat{\rho}_j = \frac{1}{N}\sum_{i=1}^{n} y_j(x_i) \tag{6.7}$$

where H is the total number of hidden layer, ϕ is the sparsity term's weight, and ρ is the sparsity parameter, which is typically very near to 0; $KL(\rho\|\hat{\rho}_j)$ is Kullback–Leibler (KL) divergence; $\hat{\rho}_j$ is the average activation of the hidden unit j.

6.2.2 *Stacked Autoencoder*

Popular research on machine learning methods has demonstrated the strong ability of deep structures in learning and their complex and nonlinear patterns [10]. These studies served as inspiration for the development of an SAE in this work, which consists of numerous autoencoders piled on top of one another, with the upper layer's input derived from the lower hidden layer's output. The SAE has a considerable advantage in extracting the dynamics' characteristics and identifying the extremely nonlinear correlations between the acquired data [11].

A conventional predictor is placed on top of the SAE model to use it for the prediction of operational condition in gas transmission pipeline systems. In this chapter, the logistic regression layer is used as a unique design in the model to predict and monitor the operating conditions of the system. The entire deep learning model, which made up of the predictor and the SAE model is shown in Fig. 6.2.

Gradient-based optimization allows the BP approach to be directly applied to the training of traditional neural networks. Deep neural networks trained using the BP approach, however, perform poorly. On the other hand, successful development of the greedy layerwise unsupervised algorithms has been made [12]. The important elements are as follows: First, the deep network was pretrained by layer using a bottom-to-top approach, using the greedy layerwise unsupervised technique; second, performance is improved by fine-tuning the model's parameters top-down on the basis of BP. The training process is displayed on Fig. 6.3.

To keep the deep learning model performing well in a changing environment, we can see that it is essential to modify and retrain the model using different new data.

The form of the evolution of operating conditions in the short-term future is not only based on the recent past but also depends on the size of the pipe network system. In order to make effective use of the information provided by the useful data, we must update the historical data for the proposed model. Therefore, the moving window method [13] is employed in this chapter. The schematic of the moving data window method is shown in Fig. 6.4.

The input and prediction portions of a moving data window's length are shown in Fig. 6.4. The inputs in each prediction step are the white areas in the diagram, and the expected future data are those referred to by the blue areas. In order to obtain better prediction performance, the overlapping parts in Prediction II are substituted with the latest predicted data received at each step shown in Prediction I.

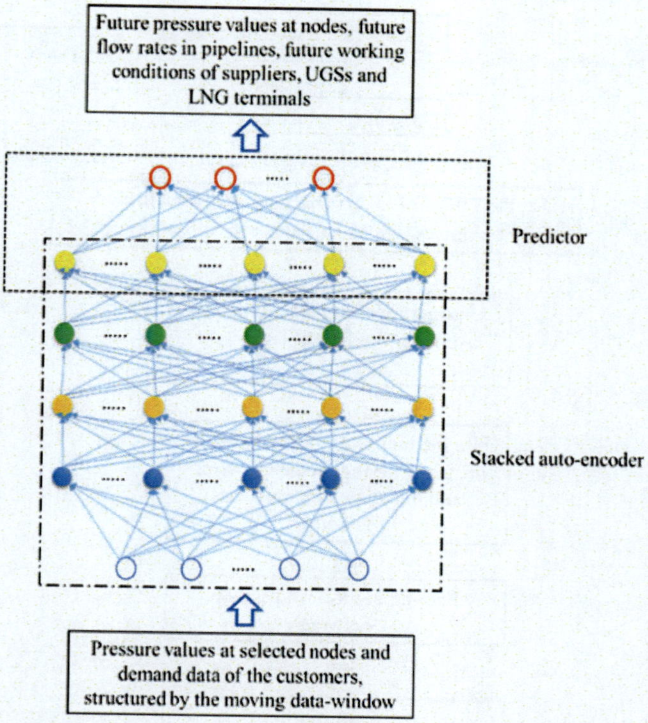

Fig. 6.2 A stacked autoencoder-based model for predicting the real-time operational status of natural gas transmission pipeline systems

A decent prediction performance can be achieved with less computational storage expense if the moving data window is designed properly. Actually, the deep learning model's "memory capacity" is determined by how long the white zone is. As a result, moving window method aids in the proposed model to effectively remember the most pertinent data for prediction. It is worth noting that this study assumes that the future state of the subject of study (the gas transmission pipeline) is influenced by past operating conditions of the system itself and by fluctuations in external gas demand, respectively. Moreover, we only consider historical needs and not generalities. The input window criteria are designed as Eq. 6.8:

$$S = \max\{T, \max\{P_1, P_2, \ldots, P_n\}\} \tag{6.8}$$

where S represents the input length; the maximum time for the disturbance to spread throughout the pipeline system is denoted by T and P_i refers to the shortest period that has an impact on the evolution of future gas demand i. By reviewing past studies in demand forecasting, demand history, natural factors, population, and social policies, all have an impact on gas demand [14].

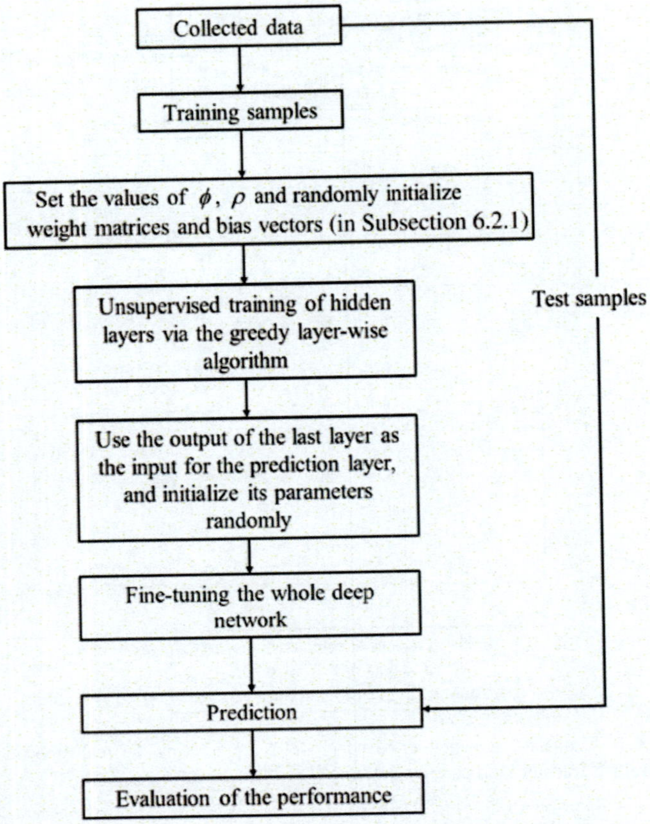

Fig. 6.3 Training framework of the proposed deep learning prediction method

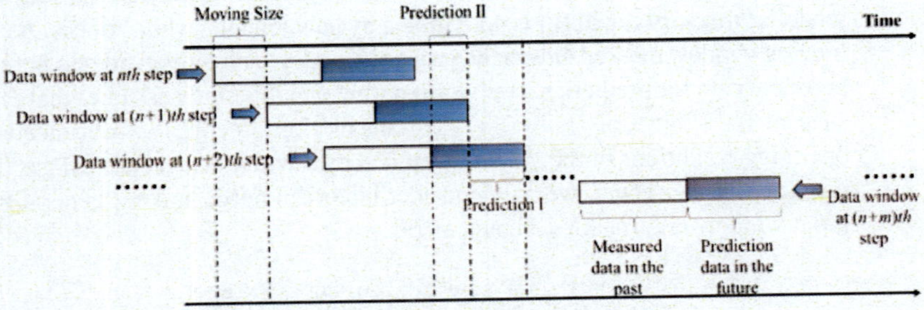

Fig. 6.4 Schematic diagram of the moving window method

6.3 Data Cleaning

Unfortunately, relying only on the information gleaned from sensors might result in a plethora of storage problems and faulty forecasts [15]. In order to narrow the scope of the issue and transform the "big data" into "wise data," it is necessary to choose relevant information from the acquired data.

Pressure is the driving force behind the pipeline system's dynamic evolution [16], and the theory of control allows for complete regulation of the system. As such, we may try to utilize the pressure data (historical and present) of these components to anticipate the state of the whole pipeline network, making the pressures of these nodes a significant source of information for system controllability.

In this chapter, the parts of the gas pipeline network that may be controlled as a whole are identified using network structural controllability theory [17]. In the actual world, this technology has been used to complicated network systems like gas pipeline networks. The applications' outcomes demonstrate the capacity to accurately identify the structural control system's driving nodes.

The pipeline network for natural gas is mostly governed by nonlinear processes; nonetheless, nonlinear systems are often fundamentally comparable to linear systems in terms of their controllability [18]. Then, the driver nodes may be isolated using the standard, moment linear dynamics:

$$\frac{\mathrm{d}x(t)}{\mathrm{d}t} = \mathbf{A}\mathbf{x}(t) + \mathbf{B}\mathbf{u}(t) \tag{6.9}$$

where $\mathbf{x}(t)$ refers to the state vector of a gas transmission network topology with N nodes at time t; matrix $\mathbf{A}$, with $N \times N$ structural dimensions, represents the interactions between individual nodes on the network connection structure; And input matrix $\mathbf{B}$ is used to identify the driving nodes controlled by the time-dependent input vector $\mathbf{u}(t)$.

The Kalman rank condition can be used as a control criterion which determines the minimum of the drive nodes in classical dynamic control theory. New ideas of complexity are required [19], however, since the traditional approach struggles to cope with intricate pipeline networks. Firstly, the constructed gas transmission pipeline network is directed, whereas traditional techniques are usually applied to undirected networks. Secondly, in order to apply the Kalman rank condition, the weights of all edges of the matrix $\mathbf{A}$ are necessary, but in practice: Even if the weights could be estimated, we would still need to compute the ranks of 2^N-1 combinations by violent search, which is not feasible for large and complex gas transmission pipeline networks.

To solve this issue, we use the analytic techniques [20] to determine the smallest possible group of driver nodes. The so-called structural controllability of a system $(\mathbf{A}, \mathbf{B})$ permits finding the free parameters in $\mathbf{A}$ and $\mathbf{B}$. Both $\mathbf{A}$ and $\mathbf{B}$ are structured matrices with zeros and free parameters as their elements. In general, a structural controllable network system is controllable for most of weight combinations, except for certain psychopathic examples with zero measurements. As a result, the problem

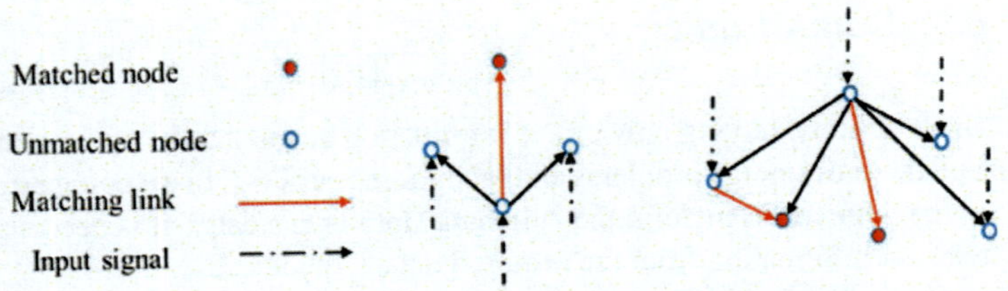

Fig. 6.5 Diagrammatic representation of the controllability of the network structure

of the components of Matrix **A** having insufficient information may be addressed with the aid of structural controllability theory.

Here, the corresponding idea from graph theory [17] is used to determine the smallest possible set of driver nodes, with the help of principles from structural controllability theory. In the directed graph, the matching set is the collection of arcs that share no vertices. As can be seen in Fig. 6.5, the node being matched needs to match the end vertex of an arc; otherwise, the node is mismatched. A pipeline network is considered fully controlled when and if all unmatched nodes are directly controlled, ensuring that the input signal can be transmitted to all matched nodes [17]. Consequently, we must determine a maximum match, which may not be unique in a given network. The maximum match is influenced by the minimum set of driving nodes, which also provides a minimum effective set for prediction. In a given directed graph, the maximum match can be found in $O(N^{1/2}L)$ steps, where L is the number of arcs in the graph.

The features of the gas transportation process and the underlying idea of structural controllability dictate that changes in pressure at the driver nodes will ultimately influence the trajectory of the whole gas pipeline network. Therefore, we can foresee how the system would develop based on the present and past pressure data of these driving nodes. This approach may drastically reduce the problem size for a complicated pipeline network while maintaining high prediction quality.

6.4 Case Study

Both the triangle-based topological pipeline network structure and the real-world portion of the gas transmission pipeline are affected by the deep learning model. Although volatility in the natural gas market is caused by nonlinear dynamics, it is chaotic and not completely random. The Mackey–Glass model is a hybrid periodic and stochastic time series model that is often used to assess the validity of forecasting models due to its chaotic nature. The Mackey–Glass model is calculated as follows [21]:

$$\frac{\mathrm{d}x(t)}{\mathrm{d}t} = \frac{ax(t-\tau_M)}{1+x^c(t-\tau_M)} - bx(t) \tag{6.10}$$

where τ_M determines the chaotic behavior of the time series and represents the time delay parameter, which must be (>16.8). Therefore, in this study, $\tau_M = 20$. The other parameters were set as follows: $a = 0.2$, $b = 0.1$, $c = 10$. Here, the time series data is created using the 4th Runge–Kutta technique, and then samples are taken at predetermined intervals. Actual data values in this unstable time series rely on those from the past, which is analogous to changes in gas consumption. In order to make the application more practical, a random term (of 1% of the nominal value of the produced gas needs) is included.

TGNET is a commercial software created by Energy Solution for both steady-state modeling of natural gas pipeline networks and for modeling the hydrodynamics of natural gas pipeline networks under transient conditions, and the simulation software is used to obtain real-time operational data of gas transmission pipelines [22]. This program has been extensively utilized in a variety of fields, including pipeline design, risk analysis, and emergency preparedness.

Although the problems encountered in real engineering are more complex and variable, this chapter focuses on deep learning models and their ability to predict the state evolution of natural gas pipeline networks with nonlinear characteristics.

The input data, together with structural stability and data-window reconstruction, dictate the size of the deep learning model's input layer. The accuracy of the results obtained after trial and error experiments using the grid search method's determines the number of nodes in the hidden layer [11]. The temperature and flow rate of the pipeline equipment as well as the gas flow rate provided by the supplier are necessary prerequisites for the prediction model.

The range of input data must be standardized within the range [0, 1] for effective learning. The normalization of each dataset was carried out independently due to the large differences between the datasets collected from different locations throughout the gas transmission pipeline system.

The exact implementation of the proposed method is shown in detail in Fig. 6.6.

The triangular pipeline network in Fig. 6.7 was used to perform an arithmetic validation, comparing the deep learning models created with two established machine learning techniques: BP neural networks and support vector machines (SVMs). In order to remove the deep learning model's inherent advantage in handling large amounts of data, we choose for the basic network. The shallow BP network just has one layer, demonstrating the superiority of the DL model over the deep neural network and the traditional machine learning model. But in order to ensure that the results can accurately reflect their capabilities, the parameters of the BP neural network and the SVM model were determined through iterative trial-and-error experiments. The provider is Node 1, and the consumers are nodes 2 and 3. Pressure control (always equal to 5 MPa) is the control mode for the supply node, while flow rate control is the control mode for the consumers. Pipelines 1–3, 1–2, and 2–3 with diameters of 0.6 m and lengths of 80 km, 90 km, and 100 km, respectively. The pipeline software TGNET default settings are used to determine the parameters and methodology for the $T-H$ simulation.

To assess the prediction performance at different prediction time lengths, this example is set to predict the change in the operating state of the pipe network after

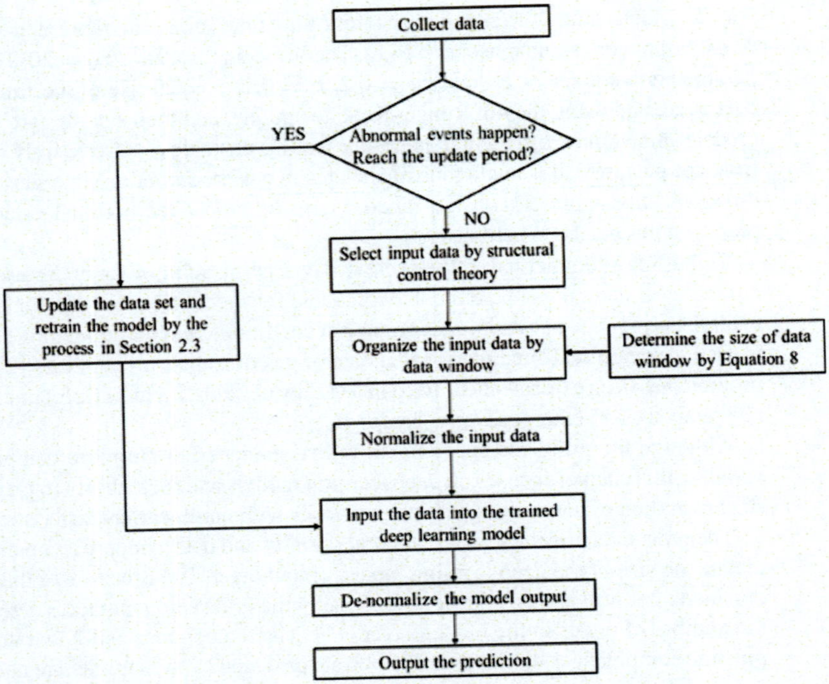

Fig. 6.6 Flowchart of the proposed method

Fig. 6.7 Topology of the triangular gas transmission network

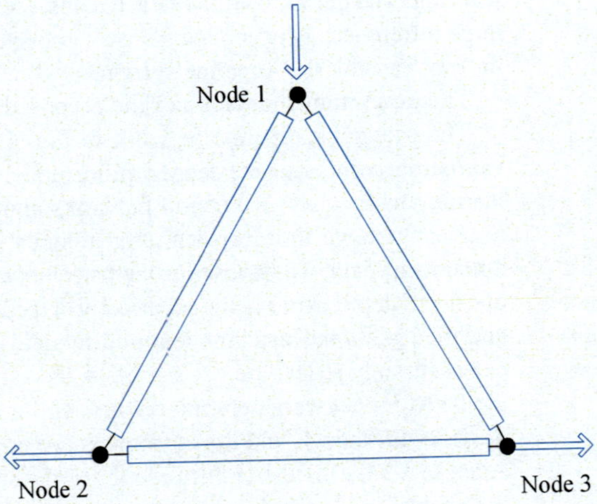

3 h, 5 h, and 9 h through the demand load and the pressure at Node 3, respectively. The length of the input time window is 10 h. Consequently, 120 is the size of the input for the triangular network's prediction issue (the length of each time window is 10 h, with samples taken every 15 min). A total of 1000 h of $T-H$ dynamic simulation was selected through the data window, and a total dataset of 4900 was used for model training, tuning, and testing, respectively, with 50% of the training set stations, 30% of the tuning data stations and the remaining 20% being the prediction set. To ensure the model is generalizable, the model is optimized and validated using several datasets. The deep learning model's parameters and structural parameters will change as prediction time increases. The sizes of the hidden layers in the "trial and error" optimization are selected from 2 to 6, and the range of the number of units in the hidden layers is [300, 250, 200, 150, 50]. The optimal architecture for the three-hour forecast is a two-layer with hidden units of 200 and 150; for the five-hour prediction, it is a two-layer with hidden units of 250 and 100; and for the ten-hour prediction, it is a three-layer with hidden units of 250, 150, and 150.

Figure 6.8 shows some of the demand generated at customer nodes 2–3. As can be seen from the figure, the generated chaotic demand data shows some degree of cyclical variation, but that variation is not completely stable, as is the case with real-world gas demand variation.

Three indices are used to evaluate the accuracy of the predictions: the mean absolute error (MAE), the root-mean-square error (RMSE), and the mean relative error (MRE). Tables 6.1 and 6.2 analyze the performances of several techniques. Table 6.1 shows the results of the pressure prediction of client Node 2, while Table 6.2 shows the outcomes of supply Node 1's gas flow rate prediction. The tables show that, for the various durations of forecast time, compared to BP and SVM, the DL model has more predictive power. We can see from Tables 6.1 and 6.2 that

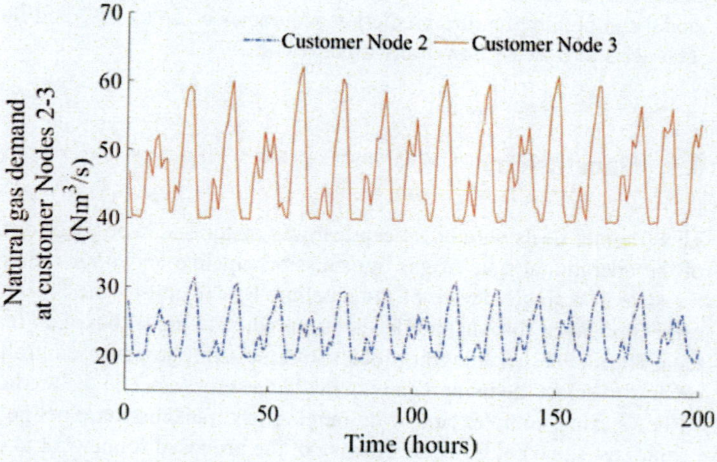

Fig. 6.8 Demand generated at customer nodes 2–3 (0–200 h)

Table 6.1 Comparison of prediction performance of different models (based on pressure prediction results for customer Node 2 (×103 Pa))

Task	The DL model		BP		SVM	
	MAE	RMSE	MAE	RMSE	MAE	RMSE
3-h prediction	2.90	3.45	12.63	17.22	17.11	22.30
5-h prediction	3.92	4.69	16.99	21.72	19.39	22.98
9-h prediction	4.58	5.67	25.78	27.23	27.18	31.10

Table 6.2 Comparison of gas flow prediction results from different models (based on supply Node 1 (Nm3/s))

Task	The DL model		BP		SVM	
	MAE	RMSE	MAE	RMSE	MAE	RMSE
3-h prediction	0.88	1.04	3.63	4.91	5.00	6.32
5-h prediction	1.15	1.37	5.06	6.43	5.73	6.75
9-h prediction	1.31	1.43	6.00	6.32	7.94	9.71

accuracy declines as forecast time increases. This is often due to the fact that when the prediction time horizon lengthens, the strength of the link between the present state and the future state weakens, making it harder for neural networks to learn such a relationship. Controlled iterative procedures that employ the predictions to solve this problem may improve the performance of the deep learning model.

Figures 6.9, 6.10 and 6.11 provide visual representations of the MRE findings in the form of empirical CDF, which depicts the change in system operating conditions forecast owing to demand randomness. As can be seen from the Figures, the DL model outperforms the BP and SVM, according to these results as well. The DL model can obtain pressure prediction accuracies of up to 99% and flow prediction accuracies of over 98% over any time period.

6.5 Conclusion

This Chapter deals with how deep learning methods can be used for the prediction of the operational state of gas transmission pipeline networks, rather than just for the state of a single device of the pipeline. By integrating the SAE model and the regression layer, this chapter develops a prediction model based on real operational data. The model simplifies the problem by employing structural controllability theory on how to select the input data with the highest relevance to the prediction.

For a fairly complex and wide-ranging gas transmission pipeline network, this chapter tests and confirms the efficacy of the proposed framework in terms of input type, prediction duration and noise level. The superiority of the proposed model is confirmed by comparing the accuracy of the proposed model with other machine

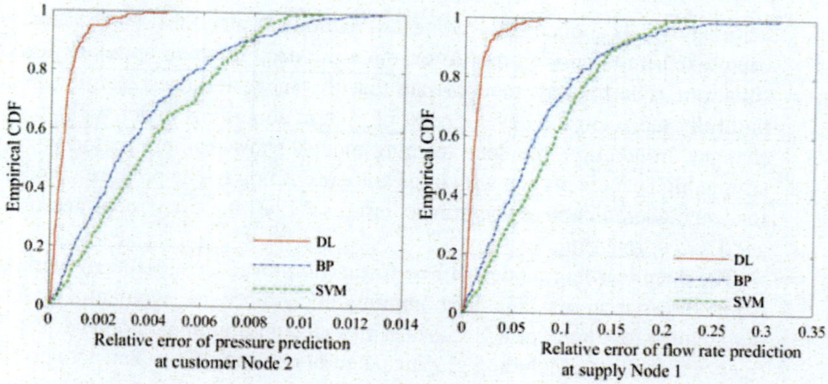

Fig. 6.9 Comparison results of the empirical CDF values of the three models (3-h predictions)

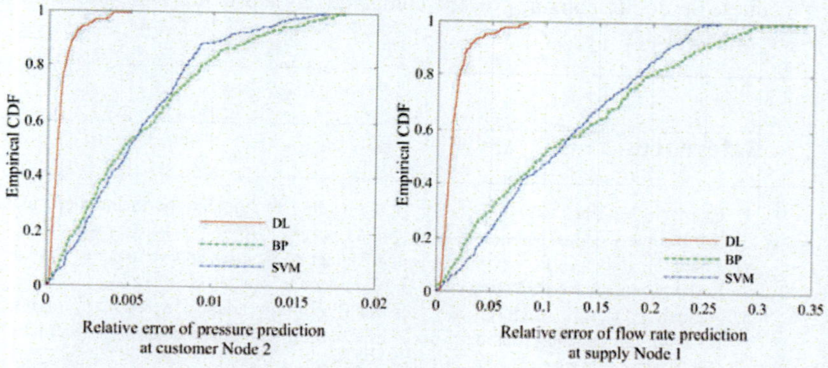

Fig. 6.10 Comparison results of the empirical CDF values of the three models (5-h prediction)

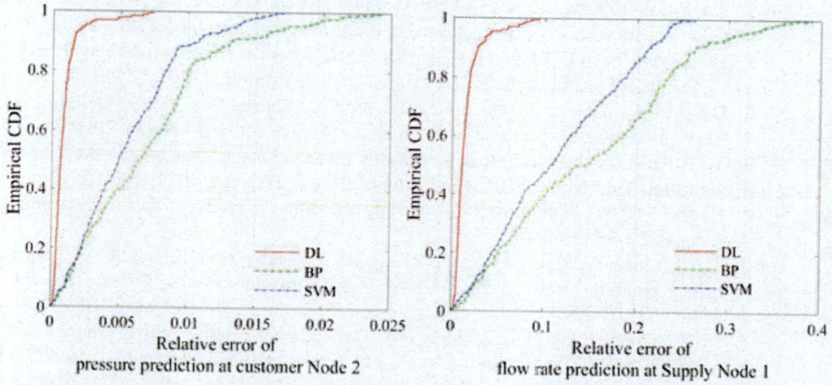

Fig. 6.11 Comparison results of the empirical CDF values of the three models (9-h prediction)

learning models applied in the context of natural gas pipelines. Three scenarios have been taken into consideration to examine the deep learning approach for atypical situations. The findings demonstrate that the suggested deep learning model can faithfully represent the development of system conditions under various abnormal changes. In addition, the deep learning models show resilient performance, maintaining high accuracy even with large amounts of noise. The proposed model allows for accurate prediction of operating pressures at LNG terminals, compressor stations, and UGS in real time.

The deep learning model will be further improved in subsequent work, such as demand response, dynamic programming, uncertainty management, etc. As part of our ongoing research, we will also take into account a number of other effective intelligence techniques, including graphical neural networks, Bayesian networks, migration networks, etc. We will compare these techniques' performance to the suggested deep learning technique. There is a need for a more efficient data-processing approach due to the de-normalizing process' comparatively substantial inaccuracies in pressure prediction.

References

1. Chertkov M, Backhaus S, Lebedev V. Cascading of fluctuations in interdependent energy infrastructures: Gas-grid coupling. Applied Energy. 2015;160:541–51.
2. Zhang D, Cadet C, Bérenguer C, Yousfi-Steiner N. Some improvements of particle filtering based prognosis for PEM fuel cells. IFAC-PapersOnLine. 2016;49:162–7.
3. Azadeh A, Gaeini Z, Haghighi SM, Nasirian B. A unique adaptive neuro fuzzy inference system for optimum decision making process in a natural gas transmission unit. Journal of Natural Gas Science and Engineering. 2016;34:472–85.
4. Lou S, Li DH, Lam JC, Chan WW. Prediction of diffuse solar irradiance using machine learning and multivariable regression. Applied energy. 2016;181:367–74.
5. Tan P, Xia J, Zhang C, Fang Q, Chen G. Modeling and reduction of NOX emissions for a 700 MW coal-fired boiler with the advanced machine learning method. Energy. 2016;94:672–9.
6. Güler NF, Übeyli ED, Güler I. Recurrent neural networks employing Lyapunov exponents for EEG signals classification. Expert systems with applications. 2005;29:506–14.
7. Durgut İ, Leblebicioğlu MK. State estimation of transient flow in gas pipelines by a Kalman filter-based estimator. Journal of Natural Gas Science and Engineering. 2016;35:189–96.
8. Yu F, Xu X. A short-term load forecasting model of natural gas based on optimized genetic algorithm and improved BP neural network. Applied Energy. 2014;134:102–13.
9. Hinton GE, Salakhutdinov RR. Reducing the dimensionality of data with neural networks. science. 2006;313:504–7.
10. Qiu J, Liang W, Zhang L, Yu X, Zhang M. The early-warning model of equipment chain in gas pipeline based on DNN-HMM. Journal of Natural Gas Science and Engineering. 2015;27:1710–22.
11. Lv Y, Duan Y, Kang W, Li Z, Wang F-Y. Traffic flow prediction with big data: a deep learning approach. IEEE Transactions on Intelligent Transportation Systems. 2014;16:865–73.
12. Bengio Y, Lamblin P, Popovici D, Larochelle H. Greedy layer-wise training of deep networks. Advances in neural information processing systems. 2006;19.
13. Zhou D, Gao F, Breaz E, Ravey A, Miraoui A. Degradation prediction of PEM fuel cell using a moving window based hybrid prognostic approach. Energy. 2017;138:1175–86.

14. Karadede Y, Ozdemir G, Aydemir E. Breeder hybrid algorithm approach for natural gas demand forecasting model. Energy. 2017;141:1269–84.
15. Xue P, Zhou Z, Fang X, Chen X, Liu L, Liu Y, et al. Fault detection and operation optimization in district heating substations based on data mining techniques. Applied energy. 2017;205:926–40. https://doi.org/10.1016/j.apenergy.2017.08.035.
16. Pambour KA, Bolado-Lavin R, Dijkema GP. An integrated transient model for simulating the operation of natural gas transport systems. Journal of Natural Gas Science and Engineering. 2016;28:672–90.
17. Leitold D, Vathy-Fogarassy Á, Abonyi J. Controllability and observability in complex networks–the effect of connection types. Scientific reports. 2017;7:1–9.
18. Chiang N-Y, Zavala VM. Large-scale optimal control of interconnected natural gas and electrical transmission systems. Applied Energy. 2016;168:226–35.
19. Zio E. Challenges in the vulnerability and risk analysis of critical infrastructures. Reliability Engineering & System Safety. 2016;152:137–50.
20. Liu Y-Y, Slotine J-J, Barabási A-L. Controllability of complex networks. nature. 2011;473:167–73.
21. Sharma V, Yang D, Walsh W, Reindl T. Short term solar irradiance forecasting using a mixed wavelet neural network. Renewable Energy. 2016;90:481–92.
22. Faertes D, Saker L, Heil L, Vieira F, Risi F, Domingues J, et al. (2010). *Reliability modelling: petrobras 2010 integrated gas supply chain.* International Pipeline Conference.

Chapter 7
Intelligent Inspection for Pipeline System

Siya Cai and Yamin Yan

Abstract It is important to monitor the pipeline on a regular basis to reduce the possibility of leaks, spills, and theft. This inspection also serves to record real occurrences and their environmental impacts. With the support of advanced GPS, sensors, and communication technologies, unmanned aerial vehicles (UAVs) provide the support of intelligent inspection for pipeline systems. UAV path planning is an important part of pipeline inspection, which is to design a flight path with the least cost or time in advance. This chapter introduces the development of mathematical models and efficient solution algorithms for UAV path planning. The model strictly considers multiple inspection missions, the intricate terrain threats, and the harsh operational requirements. Meanwhile, for the complex structure and the large scale of pipeline, improved intelligent algorithm is used to improve the efficiency of model solution.

7.1 Inspection of Oil and Gas Pipelines

Oil and gas pipelines are characterized by high pressure and flammability, sometimes accompanied by leaks, fires, and explosions [1]. Regular pipeline inspections can reduce the danger of accidents, which frequently result in significant human casualties and substantial economic expenses. Traditional manual inspection with low efficiency and high consumption has great limitations in mountainous areas, rivers, canyons, and other geographical environments, as well as in emergencies such as rainstorms, earthquakes, and landslides. Unmanned aerial vehicles (UAVs) have

S. Cai (✉)
Beijing Key Laboratory of Urban Oil and Gas Distribution Technology, China University of Petroleum-Beijing, Fuxue Road No.18, Changping 102249, Beijing, China
e-mail: 965615693@qq.com

Y. Yan
State Key Laboratory of Alternate Electrical Power System with Renewable Energy Sources, School of Renewable Energy, North China Electric Power University, Beinong Road, Changping 102206, Beijing, China
e-mail: yanym0910@163.com

H. Su et al. (eds.), *Advanced Intelligent Pipeline Management Technology*,
https://doi.org/10.1007/978-981-19-9899-7_7

drawn a lot of interest in recent years thanks to their benefits of low cost, high efficiency, and high safety. By carrying the appropriate detecting instruments for various targets, UAVs can carry out a range of duties, including leak detection, pipeline environmental condition monitoring, and pipeline infrastructure. Power grids, railroad networks, and other networks were the focus of recent study on civil UAV uses for pipeline inspection paths [2]. However, the architecture of the oil and gas pipeline networks is more intricate [3–6] and challenging to determine. Even when the best option is sought after, it is frequently impossible to do so.

There are now three ways to organize the pipeline network inspection route. The first type of algorithm is one that uses graphs, such as the Dijkstra algorithm, probabilistic roadmap, and Voronoi diagram. The inspection area will be divided into multiple spots using the Voronoi diagram. The inspection region is then divided into numerous convex surfaces by the Voronoi diagram in accordance with each inspection point. The ideal inspection path can be found if the graph contains both the initial point and the destination position. The probabilistic roadmap is used to discover the path on the roadmap by the search algorithm after converting the space into a discrete space, which can significantly increase search efficiency. In directed graphs, the shortest path problem can be resolved using the Dijkstra's algorithm. This algorithm's primary objective is to extend the outer layer, which is centered at the beginning point, all the way to the finish point. The shortest path algorithm between two vertices is the Dijkstra's algorithm. The Voronoi diagram [7, 8] and the probabilistic roadmap [9] have been used in robot path planning. Chen et al. [10] proposed a multi-objective optimization model to solve the mobile agent problem in wireless sensor networks and adopted the improved Dijkstra algorithm in the solution part of the model. However, in practice, the number of sampling points is different, which will easily lead to the deviation of the path search results, and the graph-based algorithm is difficult to be applied to the movement of UAVs.

The second strategy uses the A* algorithm and Sparse A* Search (SAS), two traditional heuristic search methods [11–13]. The state space, first developed by Hart et al. [14], is used by the A* algorithm to identify the best route from the starting location to the goal position. In order to narrow the scope of the UAV's search and enable real-time path planning, more limits are introduced to the UAV flight process. The Sparse A * Search is improved on the basis of the A* algorithm, which is pruned to avoid useless intermediate steps and nodes in the algorithm search space. However, the A* algorithm and the SAS algorithm must rely on all known environmental information to plan the path. The bigger drawback is that as the number of checkpoints increases, the search space becomes larger and the computational time of classical heuristic search algorithms increases exponentially.

The third strategy is the current heuristic search method, which is discussed in Chapter 2 and primarily consists of the genetic algorithm (GA), particle swarm optimization (PSO), and ant colony optimization (ACO). A unique method for path planning for an electrical distribution network patrol was developed by Shen et al. [15]. To determine the ideal patrol path, the vehicle routing problem (VRP) model and the modified ACO were applied. However, it is simple to enter a local optimum during the solution process for complex optimization problems, which results in premature convergence and missing the ideal answer.

Additionally, this chapter proposes a mixed integer nonlinear programming (MINLP) model based on the real requirement for pipeline network inspection. The goal of the model is to reduce the overall length of the inspection path while taking into account restrictions like the mission scenario and UAV safety performance. A two-stage solution approach is used to find the best detection path. In order to improve the efficiency and standard of path planning, an adaptive genetic simulated annealing approach (AGASA) is suggested in this chapter to solve the model for the intricate network of oil and gas pipelines.

7.2 Mathematical Model for UAV Path Planning

7.2.1 *Preliminaries*

All potential paths are assessed in the workplace by our path planner. The model uses the two-dimensional coordinates of the nodes between pipe segments as deterministic parameters and divides the pipeline with linear distribution into several pipe segments. It should be emphasized that the flight route is made up of straight segments, such as the series of segments joining the beginning node and the path node of the destination node, where the starting node is also the target node.

7.2.2 *Objective Function*

Assume that k UAVs can be used to inspect the network of oil and gas pipelines. The following definition is applied to reduce the length of the overall detection path:

$$\min F = \sum_{i}\sum_{h}\sum_{k}\left(L_{i,h}B_{i,h,k}\right) \quad i \in I; h \in H; k \in K \tag{7.1}$$

$$L_{i,h} = \sqrt{(x_i - x_h)^2 + (y_i - y_h)^2} \tag{7.2}$$

where $L_{i,h}$ indicates how far inspection node i is from node h. $B_{i,h,k}$ is a binary variable that is either 1 or 0 depending on whether UAV h executes an inspection task from node i to node h. (x_i, y_i) and (x_h, y_h) are the coordinates of node i and node h.

7.2.3 Constraints

Every UAV starts at the same location, and each UAV may only complete the inspection path once before returning to its starting location.

$$\sum_{h}^{N} B_{o,h,k} = 1 \quad h \in H; k \in K \tag{7.3}$$

$$\sum_{i}^{N} B_{i,o,k} = 1 \quad i \in I; k \in K; i \neq h \tag{7.4}$$

Each node needs to be examined by the UAV in order to guarantee the quality of the inspection.

$$\sum_{h}^{N} B_{i,h,k} = R_{i,k} \quad i \in I; h \in H; k \in K \tag{7.5}$$

The total flight time of each drone is limited by its endurance, where $R_{i,k}$ is a binary variable; if UAV k has checked node i, it is 1; otherwise, it is 0.

$$\sum_{i}\sum_{h} \alpha L_{i,h} R_{i,k} \leq E_k \quad i \in I; h \in H; k \in K \tag{7.6}$$

Set $a_i = (x_i - x_{i-1}, y_i - y_{i-1})$, and the maximum turning angle regulates the UAV's flight angle. E_k defines the endurance of UAV k, h.

$$\frac{a_i a_{i+1}^T}{|a_i||a_{i+1}|} \geq \cos(\theta) \quad i \in I \tag{7.7}$$

When completing inspection activities, UAVs must strictly follow the predetermined path and must keep a minimum safe distance d_s between one another in order to maintain the safety and order of the inspection tasks.

$$d \geq d_s \tag{7.8}$$

7.3 Two-Stage Solution Methodology

7.3.1 *First-Stage Solution*

To optimize the UAV inspection path, we provide a two-stage solution technique in this section. According to the UAV's vision, the pipeline is first separated into a few nodes.

When the UAV flying height is H and the UAV viewing angle is a, the inspection radius is $R = H \tan a$. Figure 7.1 displays a schematic design of the examination area. The pipe section is divided into a number of nodes in accordance with the detection radius. The entire pipe segment of 2R length may be seen when the UAV flies over the node. Therefore, the UAV can detect the entire pipeline by detecting each node, as shown in Fig. 7.2. In this approach, the model's size is significantly decreased, the detection task's effort is lowered, and the effectiveness of the detection process is increased.

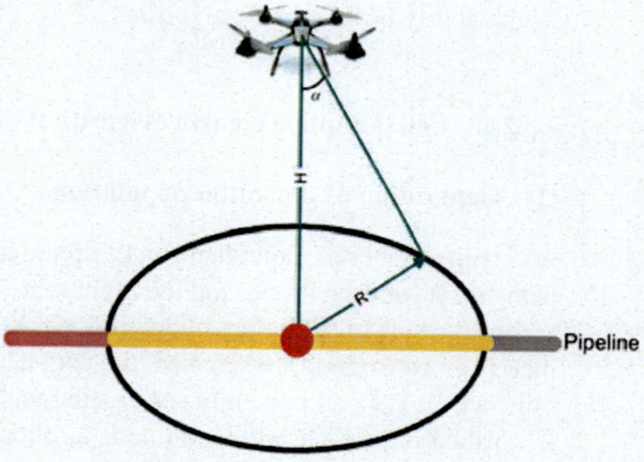

Fig. 7.1 Conceptual rendering of the inspection area

Fig. 7.2 UAV conducting inspection mission in top view

7.3.2 Second-Stage Solution

The AGASA is used in the second stage to solve the model and provide the ideal patrol route for a network of oil and gas pipelines. A type of self-adaptive heuristic algorithm known as the self-adaptive genetic simulated annealing algorithm combines the genetic algorithm (GA) with the simulated annealing technique (SA).

The "survival of the fittest" idea of biological evolution is followed by the GA as a traditional intelligent optimization method. Each GA optimization builds upon the one before it, and as new "genes" are added through the mutation procedure, the process itself is significantly altered [16]. The genetic algorithm is good at finding the global optimal solution, but it has weak local search capabilities and is prone to premature convergence. SA is a stochastic optimization method that Metropolis devised to model the heating, melting, and annealing of metals. It can eliminate local optimum solutions and make up for genetic algorithms' shortcomings. In this section, the AGASA algorithm is used to find the best solution by introducing heuristic rules in the initial population generation and crossover operation processes. This section combines the strong local search ability of the SA algorithm with the global parallel search ability of the genetic algorithm.

7.3.2.1 Self-Adaptive Genetic Algorithm

(1) **Generation of the initial population**

A certain number of individuals make up the initial population. First, a set of flight paths is set for each drone, and the flight routes start and end at the drone base. The second stage is to determine if any limits in each UAV's inspection path need to be met. If all constraints are satisfied, the path is the initial path of the UAV. If the test fails, a new inspection path is constructed until it passes. The basic scheme layout is finished once each inspection node is allocated to the UAV. What is more, the beginning path fitness is low, which helps to optimize the convergence result and essentially speeds up algorithm convergence.

Each path of the initial scheme can be coded as:

$$\begin{aligned} L_m &= (L_1, L_2, \ldots, L_k) \\ &= \begin{pmatrix} x_0, y_0, x_{11}, y_{11}, \ldots, x_{1s}, y_{1s}, x_o, y_0, x_{21}, y_{21}, \ldots \\ \ldots, x_{2t}, y_{2t}, x_0, y_0, x_{k1}, y_{k1}, \ldots, x_{kw}, y_{kw}, x_0, y_0 \end{pmatrix} \end{aligned} \tag{7.9}$$

where L_k is defined as the detection path of UAV k, x_{1s}, y_{1s} is the pipeline segment S inspected by the first UAV, and x_{2t}, y_{2t} is the pipeline segment t inspected by the second UAV. Similarly, x_{kw}, y_{kw} defines the pipeline segment w inspected by UAV k.

(2) **Fitness performance**

Fitness is a measure of a population's capacity to arrive at or come close to an ideal outcome. Higher fitness individuals are more likely to pass on to the next generation, and lower fitness persons are more likely to be removed. The model aims to determine the least value of the goal function by meeting all constraints based on the analysis mentioned above. Only four limits apply if a single UAV can perform the reconnaissance mission. All five limits should be taken into consideration if more than one UAV is needed for an inspection mission.

As a fitness function, the objective function is selected:

$$\min F = L_m = \sum_i \sum_h \alpha L_{i,h} B_{i,h,k} \tag{7.10}$$

(3) Selection of genetic operator

By replicating individuals, the selection operator creates the population of the following generation. Selection algorithms can mimic the process of biological and genetic evolution where the strongest survive. New individuals are produced by the mutation operator and the crossover operator. Both crossover and mutation operators may imitate the process of creating new biological people or species, as well as the unintentional mistake process brought on by cell division and replication.

(A) Selection operator

To choose individuals, the algorithm employs the elitist model and roulette wheel selection.

(1) Roulette wheel selection

It is possible to express the proportional relationship between the likelihood that a candidate will be chosen and their fitness value as follows:

$$P(L_m) = \frac{f(L_m)}{\sum_{m=1}^{M} f(L_m)} \tag{7.11}$$

The algorithm's precise operation is as follows:

Calculating the f_{total} by adding the values of each population member's fitness function.

In the interval $[0, f_{\text{total}}]$, f_{random} is generated at random.

Values for the fitness function rise after the first individual. If $q_m \geq f_{\text{random}}$, the chromosome L_m will be selected.

q_m is named as the accumulated probability of the chromosome L_m, and its calculation method is given in Eq. (7.12).

$$q_m = \sum_{k=1}^{M} P(L_m) \tag{7.12}$$

(2) Elitist model

The unpredictability of processes like replication, crossover, and mutation during the evolution of genetic algorithms may eliminate the members of the present population who have the highest fitness. We configure the crossover and mutation operations to exclude the people with the greatest fitness in the current population, and we utilize them to replace the individuals with the lowest fitness value following the current generation of genetic operations in order to prevent this predicament. This elite model can better help the survival of the fittest.

The elitist model's exact operational steps are as follows:

Identify the individuals in the current population with the highest fitness value f^m_{best} and the lowest fitness value f^m_{worst}.

If $f^m_{\text{best}} > f^{\text{all}}_{\text{best}}$, consider the individual with the highest fitness value f^m_{best} to be the best individuals across all populations, i.e., $f^{\text{all}}_{\text{best}} = f^m_{\text{best}}$.

Change the individual in the population with the lowest fitness value f^m_{worst} with the best individuals f^m_{best} across all populations, i.e., $f^m_{\text{worst}} = f^m_{\text{best}}$.

(B) **Crossover**

Chromosomes from two different solutions are spliced together at a crossover factor, and the spliced parts are then switched around to achieve crossover. Based on this, a novel technique involves integrating a small number of exact genes into various chromosomes to produce a whole new chromosome.

In the process of multi-segment crossover operation of the individual generated by the parent, the node on the pipe segment is not changed. This limits the UAV's ability to whiz past the pipeline and avoids repeated paths during crossing operations.

The specific steps of the crossover operation are as follows:

(1) Set the number of intersections to W and select the intersection range $p \in [p, p + W]$ randomly. Meanwhile, set $r = 1$.
(2) Check for the intersection's location in the individuals $C1$ and $C2$, and record it as x and y.
(3) Change the positions of individuals $C1(1, p)$ and $C2(1, p)$.
(4) Change the positions of individuals $C1(1, x)$ and $C2(1, y)$, $r = r + 1$.
(5) If $r < W$, proceed to Step 2; if not, the crossover procedure is finished.

$$C1 = L_m^{C1} = \left[S_1^{C1}, S_2^{C1}, \ldots, S_K^{C1}\right] = \begin{bmatrix} x_0^{C1}, y_0^{C1}, x_{11}^{C1}, y_{11}^{C1}, \ldots, x_{1s}^{C1}, y_{1s}^{C1}, x_0^{C1}, y_0^{C1}, x_{21}^{C1}, y_{21}^{C1}, \ldots, \\ x_{2t}^{C1}, y_{2t}^{C1}, x_0^{C1}, y_0^{C1}, x_{k1}^{C1}, y_{k1}^{C1}, \ldots, x_{kw}^{C1}, y_{kw}^{C1}, x_0^{C1}, y_0^{C1} \end{bmatrix} \tag{7.13}$$

$$C2 = L_m^{C2} = \left[S_1^{C2}, S_2^{C2}, \ldots, S_K^{C2}\right] = \begin{bmatrix} x_0^{C2}, y_0^{C2}, x_{11}^{C2}, y_{11}^{C2}, \ldots, x_{1s}^{C2}, y_{1s}^{C2}, x_0^{C2}, y_0^{C2}, x_{21}^{C2}, y_{21}^{C2}, \ldots, \\ x_{2t}^{C2}, y_{2t}^{C2}, x_0^{C2}, y_0^{C2}, x_{k1}^{C2}, y_{k1}^{C2}, \ldots, x_{kw}^{C2}, y_{kw}^{C2}, x_0^{C2}, y_0^{C2} \end{bmatrix} \tag{7.14}$$

Following the crossover procedure, two new individuals $D_1 D_2$ can be obtained:

$$D1 = L_m^{D1} = \left[S_1^{D1}, S_2^{D1}, \ldots, S_K^{D1}\right]$$

$$= \begin{bmatrix} x_0^{D1}, y_0^{D1}, x_{11}^{D1}, y_{11}^{D1}, \ldots, x_{1s}^{D1}, y_{1s}^{D1}, x_0^{D1}, y_0^{D1}, x_{21}^{D1}, y_{21}^{D1}, \ldots, \\ x_{2t}^{D1}, y_{2t}^{D1}, x_0^{D1}, y_0^{D1}, x_{k1}^{D1}, y_{k1}^{D1}, \ldots, x_{kw}^{D1}, y_{kw}^{D1}, x_0^{D1}, y_0^{D1} \end{bmatrix} \tag{7.15}$$

$$D2 = L_m^{D2} = \left[S_1^{D2}, S_2^{D2}, \ldots, S_K^{D2}\right]$$
$$= \begin{bmatrix} x_0^{D2}, y_0^{D2}, x_{11}^{D2}, y_{11}^{D2}, \ldots, x_{1s}^{D2}, y_{1s}^{D2}, x_0^{D2}, y_0^{D2}, x_{21}^{D2}, y_{21}^{D2}, \ldots, \\ x_{2t}^{D2}, y_{2t}^{D2}, x_0^{D2}, y_0^{D2}, x_{k1}^{D2}, y_{k1}^{D2}, \ldots, x_{kw}^{D2}, y_{kw}^{D2}, x_0^{D2}, y_0^{D2} \end{bmatrix} \tag{7.16}$$

(C) **Mutation**

A mutation is a chance adjustment to the genetic makeup. The individual to be mutated can be represented as:

$$E1 = L_m^{E1} = \left[S_1^{E1}, S_2^{E1}, \ldots, S_K^{E1}\right]$$
$$= \begin{bmatrix} x_0^{E1}, y_0^{E1}, x_{11}^{E1}, y_{11}^{E1}, \ldots, x_{1s}^{E1}, y_{1s}^{E1}, x_0^{E1}, y_0^{E1}, x_{21}^{E1}, y_{21}^{E1}, \ldots, \\ x_{2t}^{E1}, y_{2t}^{E1}, x_0^{E1}, y_0^{E1}, x_{k1}^{E1}, y_{k1}^{E1}, \ldots, x_{kw}^{E1}, y_{kw}^{E1}, x_0^{E1}, y_0^{E1} \end{bmatrix} \tag{7.17}$$

$$E2 = L_m^{E2} = \left[S_1^{E2}, S_2^{E2}, \ldots, S_K^{E2}\right]$$
$$= \begin{bmatrix} x_0^{E2}, y_0^{E2}, x_{11}^{E2}, y_{11}^{E2}, \ldots, x_{1s}^{E2}, y_{1s}^{E2}, x_0^{E2}, y_0^{E2}, x_{21}^{E2}, y_{21}^{E2}, \ldots, \\ x_{2t}^{E2}, y_{2t}^{E2}, x_0^{E2}, y_0^{E2}, x_{k1}^{E2}, y_{k1}^{E2}, \ldots, x_{kw}^{E2}, y_{kw}^{E2}, x_0^{E2}, y_0^{E2} \end{bmatrix} \tag{7.18}$$

Following are the key steps involved in a mutation operation:

(1) A random mutation probability p_m is generated in the interval [0, 1].
(2) When $0 \leq p_m \leq 0.5$, two random elements on the chromosome L_m are going to switch places, and when $0.5 \leq p_m \leq 1$, three random elements on the chromosome L_m are going to switch places.
(3) The new individuals F_1 and F_2 are generated after a mutation operation.

$$F1 = L_m^{F1} = \left[S_1^{F1}, S_2^{F1}, \ldots, S_k^{F1}\right] \tag{7.19}$$

$$F2 = L_m^{F2} = \left[S_1^{F2}, S_2^{F2}, \ldots, S_k^{F2}\right] \tag{7.20}$$

7.3.2.2 Simulated Annealing Algorithm

(1) **Initial temperature**

The initial temperature is typically set high enough to guarantee that the algorithm has strong ergodicity from the start, preventing the algorithm from entering local optima throughout the solution process. T_0 is selected as the initial temperature which can be denoted as:

$$T_0 = 10 \times N \tag{7.21}$$

(2) Temperature update function

The temperature value may be changed by the temperature update function during an external cycle. The global optimal solution must typically be reached before the temperature drops to almost zero. The equal proportional temperature update function was selected as the temperature update function in this chapter because it is efficient and simple to use:

$$T_{n+1} = wT_n, \quad n \geq 0,\ 0 \leq w \leq 1 \tag{7.22}$$

(3) Acceptance function

The acceptance function is used to determine which of the newly discovered solutions is better than the previous one. The Metropolis criterion is typically used to ascertain how well a new solution will be accepted. It is also utilized to make a distinction between the likelihood P_t that the new solution q_{new} will succeed the old solution q_{old}. t is the isothermal process's temperature. The probability is calculated using the following formula:

$$P_t = \begin{cases} 1,\ f(q_{\text{old}}) \geq f(q_{\text{new}}) \\ \exp\left(\dfrac{(f(q_{\text{old}}) - f(q_{\text{new}}))}{t}\right),\ f(q_{\text{old}}) < f(q_{\text{new}}) \end{cases} \tag{7.23}$$

7.3.2.3 Path-Planning Steps

Figure 7.3 shows the steps of UAV patrol path planning solved by AGASA algorithm, which are described as follows:

(1) Set the settings for the beginning temperature T_0, the maximum evolution generations $G_{\max}$, and the terminating temperature T_{end}. Initialize a group of pathways with the size of M.
(2) Start the genetic algebra counter $g = 0$, then assess each path's fitness.
(3) Individuals from each subpopulation were chosen based on the crossover probability and mutation probability, and after selection, crossover, and mutation processes, each new individual's fitness value was updated.
(4) Metropolis serves as the standard for determining whether to replace an old individual with a newer one.
(5) Return to Step 3 if $g < G_{\max}$, $g = g + 1$; otherwise, go to Step 6.
(6) Update the temperature and return to Step 2 if $T < T_{\text{end}}$; otherwise, the best or nearly best routes are identified.

Fig. 7.3 Second-stage solution flowchart

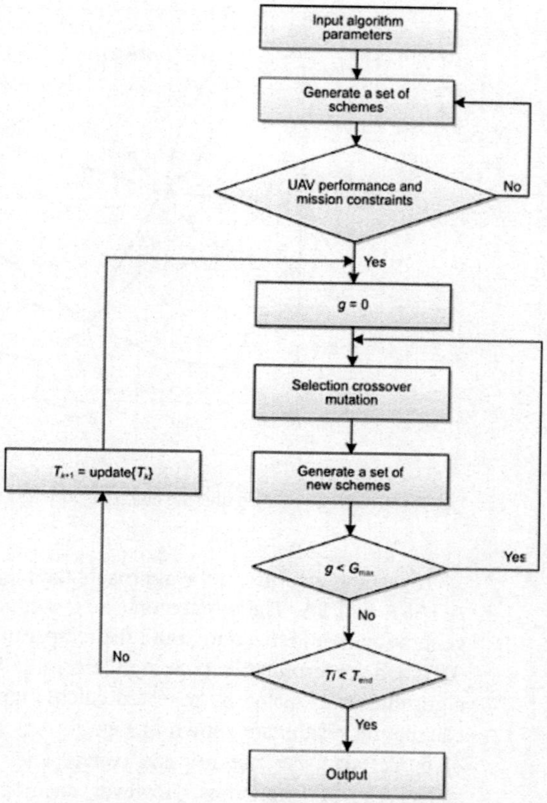

7.4 Case Study

This section takes a real case in North China as an example. The two-stage solution method proposed in this chapter builds a general framework for optimization problems and describes the concrete implementation method.

The base, which consists of a central processing facility (CPF), four valves, and twenty-nine production wells, was chosen to be constructed at the CPF. The network's topology for gathering and transit is depicted in Fig. 7.4.

Since the pipeline segment's node does not change throughout the production of the initial solution or crossover operation, the size of the model is mostly dependent on how many segments are present. For the sake of computation ease, segmentation is immediately taken into account for pipelines shorter than 1.5 km in length. Pipes longer than 1.5 km, however, will be split into parts at a 1.5 km interval. In this way, the model size can be greatly reduced, the convergence speed can be accelerated, and the optimal solution can be found quickly.

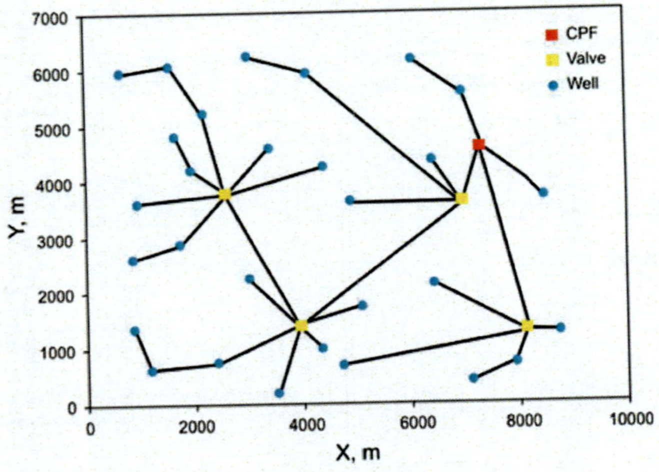

Fig. 7.4 Topology of the gathering and transportation network

To test the viability of the approach, the model was solved using two algorithms, AGASA and SA. The population size is initialized to 500, the maximum evolution generation is initialized to 3000, the temperature update coefficient is initialized to 0.97, and the termination temperature is initialized to 0.001. The stability of the two algorithms was tested by repeated calculation of the model for 10 times, and the calculation results are shown in Fig. 7.5. As shown in the table, with the increase of model size, the stability and convergence of SA are not ideal and tend to be stable after 300 iterations. However, each time the findings are inconsistent, the algorithm's calculation procedure missed the perfect solution, creating a significant danger of a local optimum. In contrast, AGASA showed good advantages in stability and convergence due to the reduction of model size and the introduction of heuristic rules. As a result, the AGASA solution suggested in this chapter may more effectively address the issue of UAV detection in oil and gas pipeline networks. The inspection pathways obtained by AGASA are depicted in Fig. 7.6, and they total 82,389 km in length.

7.5 Conclusions

For oil and gas pipeline networks, a brand-new inspection path optimization technique is put forward in this chapter. The shortest inspection path possible was chosen as the aim function in a mixed integer nonlinear programming model that also took into consideration the mission environment and the UAVs' safety performance. For the part on solving models, we recommend a two-stage solution method. In the first

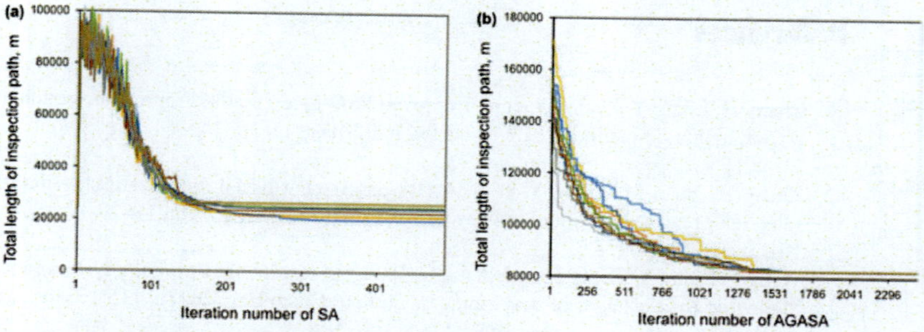

Fig. 7.5 Results of SA and AGASA's calculations

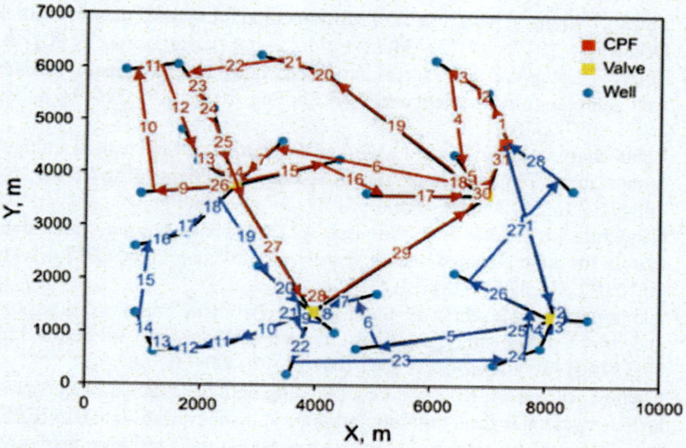

Fig. 7.6 UAV inspection route for the network of oil and gas pipelines (the first and second UAVs are indicated by the pathways in brown and blue)

step, the pipeline is split into multiple nodes dependent on the UAV's visual capabilities. The model is solved using AGASA, which incorporates heuristic rules, in the second step. Finally, the model is solved by combining the two methods SA and AGASA. The outcomes demonstrate the suggested AGASA algorithm's strong convergence and stability.

The majority of oil and gas networks may have their inspection path planning optimized using the technique described in this chapter. It should be noted that this chapter does not take into account the variations in inspection between high-leakage risk locations, highly inhabited areas, and unpopulated areas and instead assumes that each inspection node is equally important.

References

1. Zhang H, Liang Y, Zhang W, et al. Improved PSO-based method for leak detection and localization in liquid pipelines. IEEE Trans Ind Inf. 2018a; 14(7):3143–54. https://doi.org/10.1109/TII.2018.2794987.
2. Zhang H, Yuan M, Liang Y, et al. A risk assessment based optimization method for route selection of hazardous liquid railway network. Saf Sci. 2018b; 110:217–29. https://doi.org/10.1016/j.ssci.2018.04.003.
3. Wang B, Liang Y, Zheng J, et al. A methodology to restructure a pipeline system for an oilfield in the mid to late stages of development. Comput Chem Eng. 2018a; 115:133–40. https://doi.org/10.1016/j.compchemeng.2018.04.008.
4. Wang B, Yuan M, Zhang H, et al. An MILP model for optimal design of multi-period natural gas transmission network. Chem Eng Res Des. 2018b; 129:122–31. https://doi.org/10.1016/j.cherd.2017.11.001.
5. Zhang H, Liang Y, Ma J, et al. An MILP method for optimal offshore oilfield gathering system. Ocean Eng. 2017b; 141:25–34. https://doi.org/10.1016/j.oceaneng.2017.06.011.
6. Zhang H, Liang Y, Zhang W, et al. A unified MILP model for topological structure of production well gathering pipeline network. J Pet Sci Eng. 2017d; 152:284–93. https://doi.org/10.1016/j.petrol.2017.03.016.
7. Bhattacharya P, Gavrilova ML. Voronoi diagram in optimal path planning. In: 4th International Symposium on Voronoi Diagrams in Science and Engineering (ISVD 2007); 2007. p. 38–47. https://doi.org/10.1109/isvd.2007.43.
8. Candeloro M, Lekkas AM, Sørensen AJ. A Voronoi-diagram-based dynamic path-planning system for underactuated marine vessels. Control Eng Pract. 2017; 61:41–54. https://doi.org/10.1016/j.conengprac.2017.01.007.
9. Akbaripour H, Masehian E. Semi-lazy probabilistic roadmap: a parameter-tuned, resilient and robust path planning method for manipulator robots. Int J Adv Manuf Technol. 2017; 89(5):1401–30. https://doi.org/10.1007/s00170-016-9074-6.
10. Chen C, Rickert M, Knoll A. Path planning with orientation-aware space exploration guided heuristic search for autonomous parking and maneuvering. In: 2015 IEEE Intelligent Vehicles Symposium (IV); 2015. p. 1148–53. https://doi.org/10.1109/ivs.2015.7225838.
11. Gammell JD, Srinivasa SS, Barfoot TD. Batch Informed Trees (BIT*): sampling-based optimal planning via the heuristically guided search of implicit random geometric graphs. In: Proceedings of 2015 IEEE International Conference on Robotics and Automation (ICRA). Seattle, W A, USA; 2015. p. 3067–74. https://doi.org/10.1109/icra.2015.7139620.
12. Li J, Huang Y, Xu Z, et al. Path planning of UAV based on hierarchical genetic algorithm with optimized search region. In: Proceedings of 2017 13th IEEE International Conference on Control & Automation (ICCA). Ohrid, Macedonia; 2017. p. 1033–8. https://doi.org/10.1109/icca.2017.8003203.
13. Yu J, LaValle SM. Optimal multirobot path planning on graphs: complete algorithms and effective heuristics. IEEE Trans Robot. 2016; 32(5):1163–77. https://doi.org/10.1109/TRO.2016.2593448.
14. Hart PE, Nilsson NJ, Raphael B. A formal basis for the heuristic determination of minimum cost paths. IEEE Trans Syst Sci Cybern. 1968; 4(2):100–7. https://doi.org/10.1109/TSSC.1968.300136.
15. Shen X, Sang J, Sun Y, et al. Application of improved ant colony algorithm in distribution network patrol route planning. In: Proceedings of 2016 7th IEEE International Conference on Software Engineering and Service Science (ICSESS). Beijing, China; 2016. p. 560–63. https://doi.org/10.1109/icsess.2016.7883132.
16. Hua Yu , Zheyi Yao. Focusing through disturbed multimode optical fiber based on self-adaptive genetic algorithm. In: Optik-International Journal for Light and Electron Optics; 2022(261), 169129. https://doi.org/10.1016/j.ijleo.2022.169129.

Chapter 8
Probabilistic Safety Analysis in Complex Pipeline Systems

Zhaoming Yang and Huai Su

Abstract The sustainable and reliable supply of natural gas has become a critical global worry about economic, political, and technological causes because of the realization of the significance of networks of natural gas pipeline for energy security. However, a number of variables, including stochastic demand fluctuations, unexpected changes in supply capacity, and random pipeline failures, affect the oil and gas pipeline network's supply capacity. People must therefore evaluate supply security from a probabilistic rather than deterministic standpoint. This chapter introduces a methodical approach for evaluating natural gas pipeline networks' supply reliability. The created technique combines stochastic processes, graph theory, and thermal–hydraulic modeling while taking complexity and uncertainty into consideration. The case study shows that a network of gas pipelines is taken into account, and the outcomes are thoroughly studied.

8.1 Evaluation of Natural Gas Pipeline Networks for Supply Reliability

Networks of natural gas pipelines play a crucial role in coordinating supply and demand for the fuel. In the past ten years, there have been a number of unforeseen natural gas supply outages, which have had a significant impact on global economic and social stability [1]. The dependable and steady supply of natural gas has evolved into a critical global concern for political and technological causes. This illustrates the significance of the natural gas pipeline grid to energy safety.

The chance of a system, subsystem, or component "to fulfill a necessary features in defined conditions and for a stated amount of time" [2] is the most frequently

Z. Yang (✉) · H. Su
National Engineering Laboratory for Pipeline Safety; Beijing Key Laboratory of Urban Oiland Gas Distribution Technology, China University of Petroleum, 102249 Beijing, China
e-mail: yangzhaoming.upc@foxmail.com

H. Su
e-mail: suhuai@cup.edu.cn

H. Su et al. (eds.), *Advanced Intelligent Pipeline Management Technology*,
https://doi.org/10.1007/978-981-19-9899-7_8

recognized qualitative definition of reliability. However, we apply the supply function of the network of natural gas pipelines as judged by the reliability criterion for large infrastructure systems.

System reliability estimates have traditionally been calculated using system reliability analysis techniques like logic modeling and system deconstruction. When pertinent historical data are available, statistics is utilized to generate the reliability index [3].

Traditional reliability theory is unable to account for the complexity (structural and dynamic) of massive infrastructures that span vast geographic areas, run under a variety of situations, and contain a sizable number of heterogeneous components [4].

There are a wide variety of ways for simulating the random properties of complex infrastructures, including railway systems, electrical grids, and natural gas pipeline networks. To represent a complex system with uncertainties, stochastic techniques, such as Monte Carlo-based methods, Markov process-based strategies [5–8], are frequently utilized. To describe the interdependencies among essential infrastructures and the global repercussions of various scenarios, probabilistic dynamic modeling is used [9].

Recently, a powerful method for modeling transportation infrastructure systems as graphs and analyzing connection aspects has been suggested. This method research on vulnerability and reliability using graph theory. Models of the structure and function make up a structural-function model. The function model represents how the system behaves, whereas the structural model explains how the physical items in the system are represented as edges and nodes. The functional modeling might be simplified and abstracted, like flow-based methodologies, or it could be engineering-oriented, like the function model for connecting the gas supply to the grid [10].

In several studies, such as GEMFLOW [11] and MC-GENGERCIS [12], the significance of taking into account a gas supply system's operation, function, capacity, and limitations is also stressed.

The created framework combines approaches for tackling the issues from several angles, including environmental, functional constraint, topological, and dynamic. The technique can help engineers and management estimate safety margins to provide customers with comprehensive solution from a practical standpoint.

8.2 Mathematical Model for Reliability Evaluation

The established method is broken down into three sections for ease of understanding: system modeling, unit failure analysis, and reliability evaluation. Figure 8.1 depicts the framework, and the next chapter is structured as follows.

In unit failure analysis, two approaches are used: both the previous data-based approach and the failure mechanism-based approach. Pipeline failure probabilities

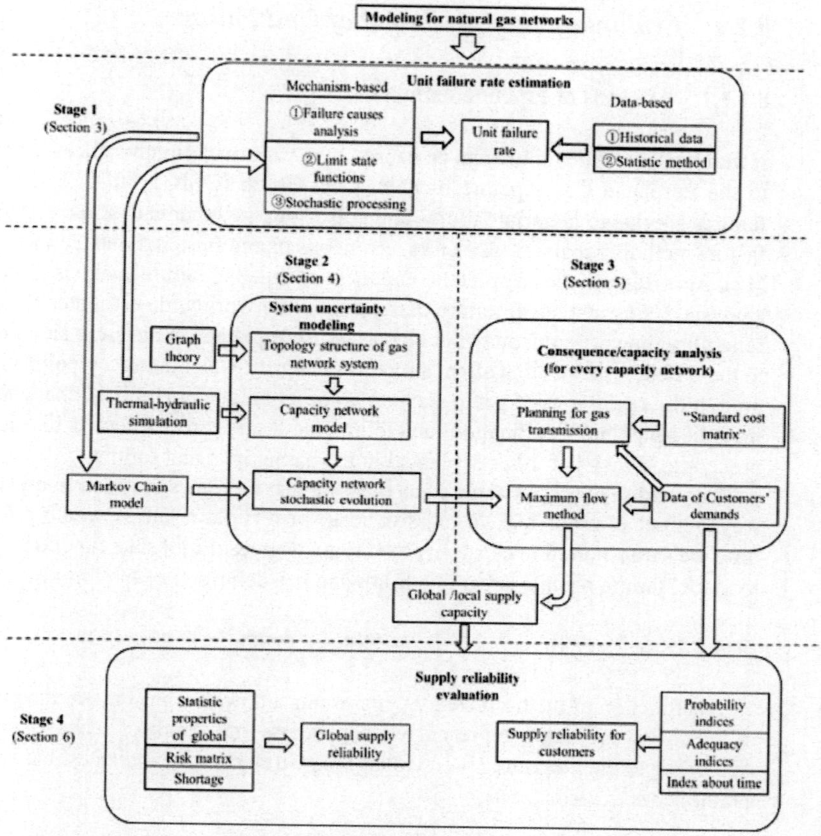

Fig. 8.1 Developed methodology's framework

vary with respect to both space and time. The others' percentage, on the other hand, is steady since the environment in which they function has little variation.

Several techniques have been developed to mimic the effects of unit stochastic failures in the natural gas industry delivery networks. This chapter simulates stochastic changes in the pipeline network system and evaluates the implications while taking complexity and uncertainty into consideration. Stochastic processes, graph theory, and thermal–hydraulic modeling are some of the techniques used.

8.2.1 *Evaluating the Likelihood of Unit Failure*

8.2.1.1 Analysis of Pipeline Failures

Natural gas pipeline failure can be caused by a variety of circumstances. According to the European Gas Pipeline Incident Data Group (EGIG), 30–40% of pipeline failures are due to material failures and corrosion; the other causes include external factors including maintenance tasks, erroneous operations, and outside intervention [13]. According to this, a crucial component of pipeline failure analysis is corrosion analysis. Using statistical failure data can provide worldwide estimates of pipeline reliability; however, this may not be correct for pipelines in different situations due to the different factors that affect structural integrity. Calculating a pipeline's failure probability requires taking into account all of its unique operational characteristics, including pipeline specifications and failure modes. Figure 8.2 depicts the analytical procedure, which is further explained in the paragraphs that follow.

A corrosion occurs in natural gas pipeline that has flaws may fail in one of three ways, either a minor leak, a massive leak, or a rupture, all of which reduce the pipeline's transportation capability to varying degrees. Limit state functions are used to model failure mechanisms. The following is a definition of the limit functions:

$$L_1 = 0.8w_t - \text{depth} \tag{8.1}$$

The function of the limit state L_1 represents a flaw that has broken the pipe wall, in which w_t and *depth* represent wall thickness and depth of leakage, severally. According to the literature [14], standard industrial practice indicates that whenever a fault depth exceeds $0.8w_t$,

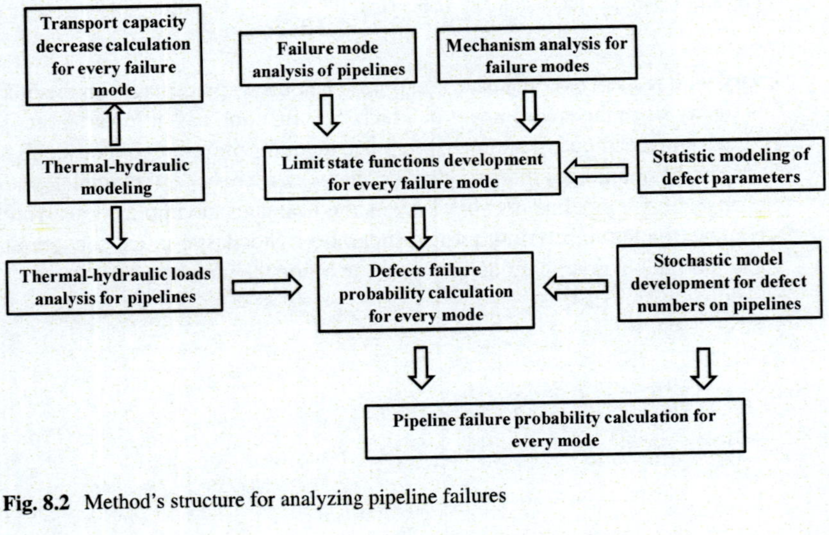

Fig. 8.2 Method's structure for analyzing pipeline failures

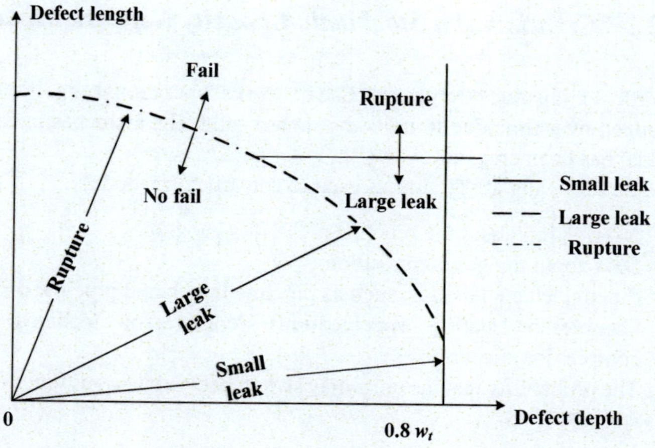

Fig. 8.3 The schematic of the distinct failure modes

$$L_2 = P_{\text{burst}} - P_{\text{in}} \tag{8.2}$$

where P_{burst} provides the corrosion defect's rupture pressure under internal pipeline pressure P_{in}. The limit state of a pipe break with plastic collapsing as a result of internal pressure is defined by this limit state function,

$$L_3 = P_{\text{rupture}} - P_{\text{in}} \tag{8.3}$$

where P_{rupture} is the pipe rupture pressure with an indeterminate extension of the defect in the axial direction.

Figure 8.3 displays a schematic of the three different failure mechanisms: small leak—$(L_1 \leq 0) \cap (L_2 > 0)$, large leak—$(L_1 > 0) \cap (L_2 \leq 0) \cap (L_3 > 0)$, and rupture—$(L_1 > 0) \cap (L_2 \leq 0) \cap (L_3 \leq 0)$ on the basis of which Monte Carlo simulation may be used to determine the likelihood of each failure mode and to determine the inner pressure P_{in}. TGNET is an offline simulation program for natural gas pipelines, used to model the thermal–hydraulic nature of natural gas transmission pipelines.

The probability of failure due to various mutually exclusive causes such as corrosion, misuse, external disturbances, and maintenance activities is summed to finally assess the pipeline failure probability.

8.2.1.2 Estimation of the Station Failure Rate

LNG ports, compressor stations, and natural gas storage facilities all operate under relatively steady circumstances as compared to pipelines. Therefore, the statistics of historical data are used to determine their failure probability.

8.2.2 *Creation of a Stochastic Capacity Network Model*

In order to illustrate how gas pipeline networks' operational conditions stochastically changed over time due to units that failed suddenly, a stochastic capacity network model has been created.

The following are the inputs used to construct the model:

a. Information about the pipeline network's geography.
b. Data about the gas composition.
c. Pipeline characteristics, such as pipeline length and pipeline diameter.
d. Compressor station specifications: compressor technical specifications, compressor allocation.
e. The probability that the units may switch between various functioning and failure states.

Figure 8.4 depicts the model creation process:

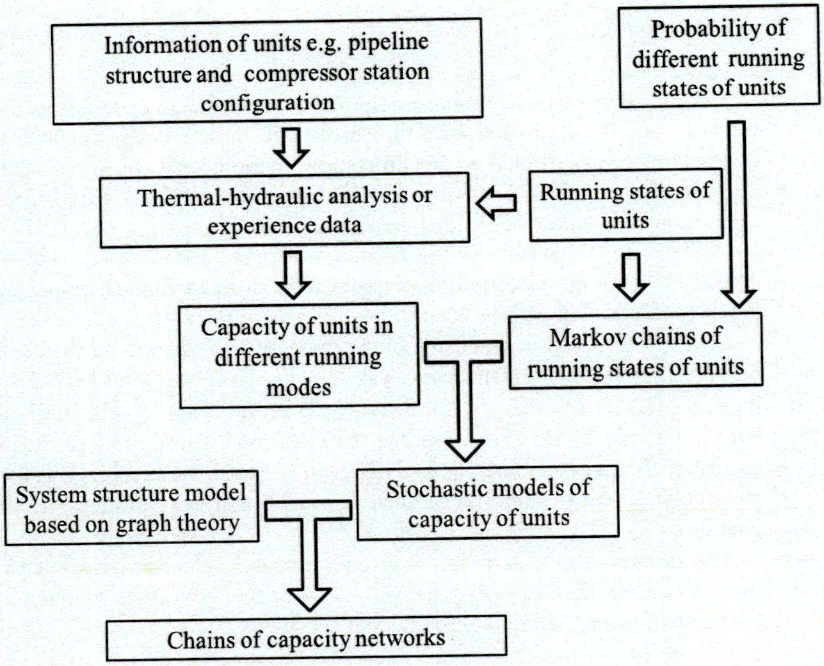

Fig. 8.4 Modeling stochastic capacity network development process

8.2.2.1 Capacity Network Model

To explain the structure of the system and the current status of its transportation capacity, a capacity network model is constructed in this case. The capacity network model represents pipelines as arcs connecting nodes that represent compressors, LNG terminals, gas storage facilities, and demanding sites. The directions of gas flows are established, and the capacity network model is guided.

Pipeline studio—TGNET's thermal–hydraulic model is used to determine the capacity weights, which indicate the transit capacity of pipelines (from Energy Solutions). Pipeline Studio is a business software and engineering solution for pipeline management design that includes graphical settings, analysis tools, as well as a simulation engine with performance that has been shown in the field.

8.2.2.2 Models for Natural Gas Pipeline Network Units' Stochastic Capacity State Transitions Based on Markov Chains

The capacity states of the network units' random variations are modeled using Monte Carlo simulation (MCS). The status probabilities are ultimately determined by solving an equational system, and the state transition rates are formally stated as probability matrices. This is what we mean when we talk about a Markov process.

For further details on the Markov process and Monte Carlo simulation, see the flowchart in Fig. 8.5:

(1) Starting time is set $t = 0$, mission time $T = 1$ year and time step $\Delta t = 1$ month.
(2) The units are first configured to their initial conditions. On account of the initial conditions, the capacity network model at $t = 0$ is produced.
(3) Based on the current states and the transition probability matrices, sample the unit states up to task time T in the interval $t = t + \Delta t$.
(4) The capacity network chains for this simulation year should be recorded. A chain of capacity network states that take place during the course of the simulation is referred to as a capacity network.
(5) Duplicate Steps (2) to (5) for Nm times ($Nm = 10^6$).
(6) The output for the capacity network model chains is required.

Pipelines can exist in one of three states, each of which corresponds to a different capacity: normal, degeneration (little or major leak), and interruption. The normal state has the option of bi-univocally changing or remaining constant with the other states. The states of interruption are returned to their starting levels in the subsequent simulation time trial. Probability of failure, the connection stated above (shown in Fig. 8.6), and historical data are used to construct the transition probability matrices of pipelines.

According to practical experience, two sorts of unexpected occurrences are set. In addition to the normal operating state, there may also be interruptions due to events like complete station failure and degeneration due to compressor failure, which are

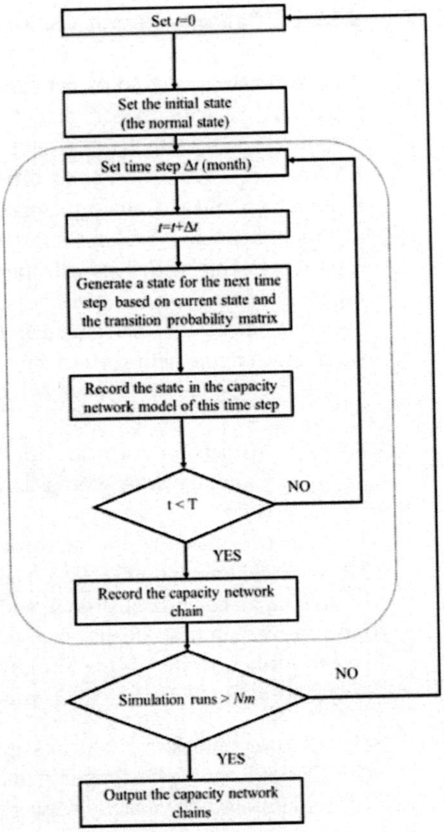

Fig. 8.5 Flowchart for the discrete Markov process development of a single unit in the pipeline network in MC simulation

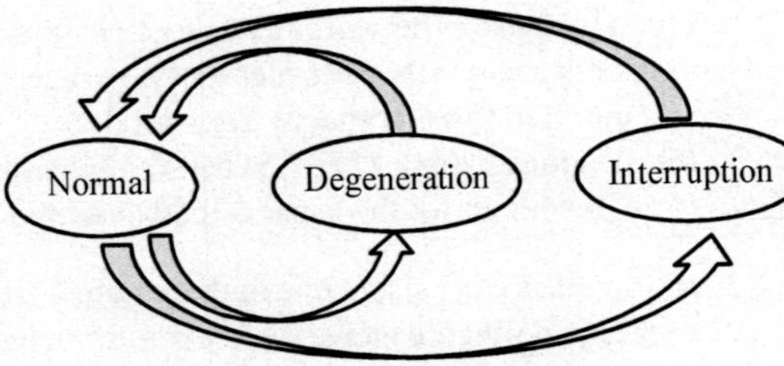

Fig. 8.6 States and pipeline state transition rules

also referred to as such in this work. Understanding the "compressor failure" incidence is simple: just the compressors malfunction, while all other station components function normally. For another, when degeneration occurs, the compressor station will still be able to transport gas, but the capacity of the neighboring pipes will be lowered to a specified level, which may be calculated by thermal–hydraulic modeling.

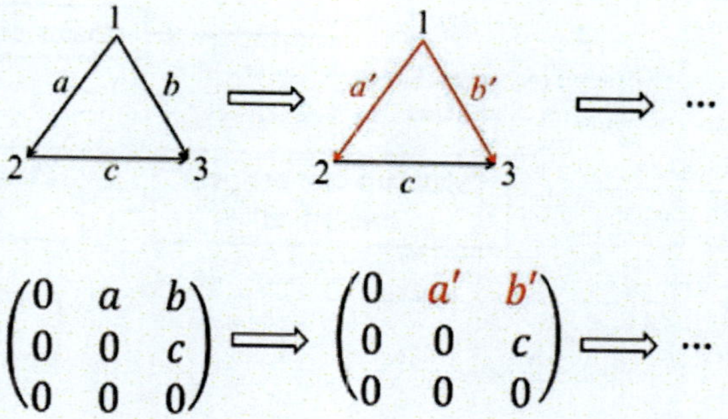

Fig. 8.7 Illustration of a stochastic network model

The ability to pressurize and transfer is both zero at the interruption stage. Incorrect operation and maintenance work are only two examples of the many causes of interruption. Again, the transitions between compressor station statuses apply in Fig. 8.5.

On the basis of the defined stochastic models, the weights of the arcs in the capacity network model change stochastically over time. The stochastic flow network model combines the capacity network model and the units of stochastic model in order to depict the dynamics of the natural gas pipeline network from a systemic viewpoint.

A straightforward illustration of the capacity network model and the system's stochastic temporal development is shown in Fig. 8.6. In Fig. 8.7, a, b, and c (a', b', c') are the arcs weights, which stand in for the pipelines' capacity.

8.2.3 Establishment of the Methodology for Capacity Analysis

The network transmission capacity will decrease as a result of stochastic unit failure, resulting in a deficit of supply for consumers. The capacity network model is used to calculate the consequences. Every Markov chain time step requires a new computation since the capacity network's constituents vary stochastically. Because the stochastic simulation uses a rather lengthy time step, transient conditions are not taken into account here (month).

The computation of the consequences involves two phases:

a. Planning for the pipeline network's natural gas transmission direction.
b. Maximizing the transmission capacity.

Figure 8.8 depicts the two components of the modeling process to illustrate how stochastic analysis and consequence analysis relate to one another in this study.

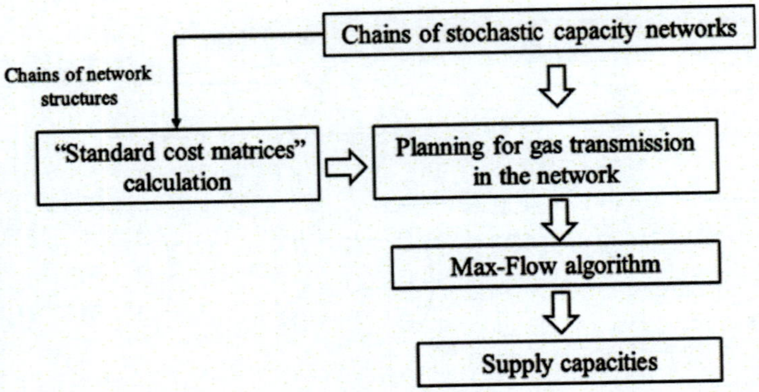

Fig. 8.8 Process consequence and gas network system capacity analysis

8.2.3.1 Building the Network's Natural Gas Transmission Path

In this chapter, we argue that the most important factors in gas transmission planning are cost effectiveness and supply distance. Political factors and societal repercussions, for example, are not taken into account in this chapter. Here is how it works.

A. **"Standard cost matrix" calculation**

The "standard cost matrix" is an adjacent matrix built using the pipeline network structure and the "standard cost" on each arc. The length of the pipeline and the cost of transmission affect what is known as the "standard cost," which is computed using Eq. (8.4). To distinguish the relative significance of these two factors, weights of importance of these characteristics are applied for the computation of "standard cost":

$$C_{ij} = \alpha L_{ij} \times \beta(Q_{ij} \times c) \tag{8.4}$$

where L_{ij} is the length of the pipe from node i to node j (km); Q_{ij} is the expected volume of gas transported from node i to node j (MCM); c is the cost of natural gas transportation (\$/(km·MCM)); α and β are the importance weights of distance and cost of transmission, respectively; C_{ij} is the optimization factor including cost and distance (\$). Even while the "standard cost" is \$, it is not the same as the real cost.

The proposed "standard cost" is more suited to current management concepts under various circumstances as compared to real cost. The cost is not usually the most crucial aspect to take into account in the actual planning process for gas transmission. Due to this, we decided to use the "standard cost" as the optimization's goal, and the various management concepts are implemented by varying the weights in Eq. (8.4).

B. **Determining the direction of gas flow**

With the exception of other sources, the "optimal pathways" from each gas source to every other node in the network are determined for transmission planning using the Floyd algorithm and the "standard cost matrix." A graph theory method called the Floyd algorithm is used to search "the shortest route" between different nodes, where the definition of "shortest" depends on how the weights on the graph's arcs are determined. Our "shortest way" calculation considers the lowest "standard cost," and as a result, it reflects the "optimal path" taking into account both transmission distance and cost.

C. **Gas transmission planning**

In this section, it established what kind of clients a gas source can serve. The closest demand sites with the best economic efficiency will receive preference, per step B, according to a source. The program then determines if the source's capacity has been reached: The algorithm will look for the next unmet demand location in the sequence discovered at step B if there is still unutilized capacity and unhappy consumers. Until the source's remaining capacity is zero or all clients are content, this procedure is continued.

D. **Adjustment under different conditions**

The "standard cost matrix" will alter along with the capacity network as it stochastically changes over time. As a result, following the unforeseen occurrences, the gas transmission task in the network should be changed. The procedures from A to C are carried out at every time step in order to react to the present network environment because these occurrences happen at random times.

8.2.3.2 Transmission Capacity Calculation

Finding the optimal distribution strategy becomes more challenging for large-scale and complicated networks. Resource distribution frequently involves the application of game theory and operational research techniques. Additionally, graph theory has become a useful alternative when taking into account the necessity of computing efficiency.

In graph theory, the computation of supply capacity under unforeseen circumstances is transformed into a maximum flow issue. Calculating the greatest capacity of transmission between two nodes in a transportation network is the goal of a maximum flow issue (source and sink). The maximum flow problem has been tackled in graph theory using a variety of methods, including the Ford-Fulkerson algorithm, the Dinic algorithm, and others. The Ford-Fulkerson method is applied in two steps:

- Search for edges with capacity that connects the source and sink.
- Continue searching until no more flows can be added to the path.

Two restrictions should be adhered to during the search:

- A node's entire flow into and out of it must equal one another.
- The edge flow does not go in the wrong direction and stays within permitted limits.

In a gas network, there are many sources and sinks, and any flow from a source can be transferred to any sink. In graph theory, this is referred to as a multiple sources and sinks issue. This issue may be simplified to a one-source, one-sink problem by supposing a "super source" and "super sink" connecting with every source and sink by edges of unlimited capacity.

8.2.4 Supply Reliability Assessment of Natural Gas Pipeline Network

A natural gas network's supply reliability is evaluated on both a global and an individual aspect. The individual aspect indicates the network system's capacity to meet consumer needs, while the global aspect represents the network system's functional integrity. The former factor is more significant when assessing supply reliability. Customers might have different demands, which is a critical part of supply security.

Similar to what is done for electrical grids, indices are constructed from three factors: time, sufficiency, and probability. The frequency of each customer's displeasure is estimated using reporting probability indices (using Eqs. (8.5) and (8.6)), which are also used to predict how well the gas network will be able to please each customer in the case of unanticipated situations. Adequacy is used to record the effects of unforeseen occurrences on the system and each individual consumer. The created indices are meant to aid in quantifying supply security management concepts.

A. Indices of probability

In pipeline supply networks, reliability might be determined differently for various clients. In this example, the rate of shortage (ROS), which is similar to the traditional failure rate, is used to show the reliability of supply at a demand site using an exponential estimation of Eq. (8.5). Equation (8.6) provides an estimate of *ROS.*

$$\text{Reliability} = \mathrm{e}^{-\text{ROS}\cdot t} \tag{8.5}$$

$$\text{ROS} = \frac{\text{Number of customer shortages}}{\text{number of months}} \tag{8.6}$$

B. Measures of adequacy

$$\text{Global average natural gas shortage} = \frac{\text{sum of all global natural gas shortages}}{\text{number of months}} \tag{8.7}$$

Global average natural gas shortage in Eq. (8.7) refers to the overall likely lacking in gas provided by the gas pipeline system in one month. The total quantity of worldwide gas shortages is the total amount of gas shortages that were simulated.

$$\text{Average number of unsatisfied customers} = \frac{\text{sum of unsatisfied customers}}{\text{number of months}} \quad (8.8)$$

Average number of unsatisfied customers is defined in Eq. (8.8) as the monthly average of consumers whose requests are not entirely met by the pipeline network. The term "*sum of unhappy customers*" refers to the total number of disappointed nodes (or customers) throughout all simulations.

$$\begin{aligned} &\text{Average natural gas shortage of a customer} \\ &= \frac{\text{sum of natural gas shortage of a customer}}{\text{number of months}} \end{aligned} \quad (8.9)$$

In Eq. (8.9), "*average natural gas of a customer*" refers to the average monthly natural gas shortfall for one customer. The total quantity of natural gas shortage by one customer during the simulations is referred to as the "*sum of natural gas shortage of a customer*."

C. Index of duration of shortage

$$\text{Average shortage duration} = \frac{\sum_{\text{times of shortage of the customer}} \text{duration of each shortage event}}{\text{number of months}} \quad (8.10)$$

The length of the shortage event is sampled at random in Eq. (8.10). The average length of a customer's monthly shortage situation is referred to as *average shortage duration.*

8.3 Case Study

An imaginary natural gas network has been used to test the applicability of the suggested approach. That assume accurate and consistent information about the pipeline system, such as pipeline specifications, client requests, and ability to deliver gas. Figure 8.9 displays the network.

Monitoring the operating point and making sure it is contained within the storage envelope allow for the real using the natural gas storage facility. However, the LNG functioning stock and regeneration capacity place a restriction on the LNG terminal's operational area.

Thermal–hydraulic simulation was used to determine the transportation capabilities of pipelines under unanticipated conditions. Pipeline's thermal–hydraulic

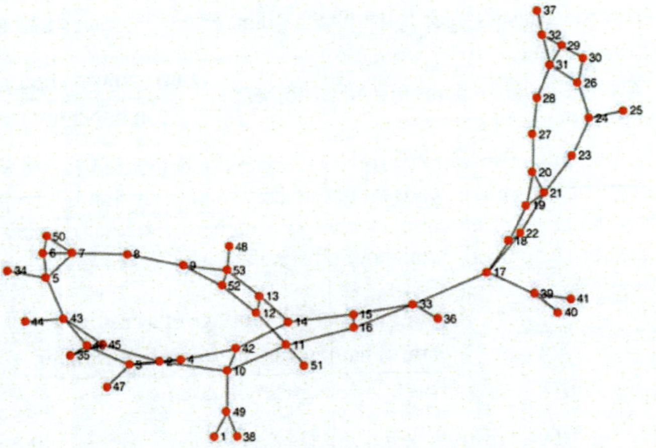

Fig. 8.9 Network design for natural gas transmission

models were created using TGNET's professional software with the recommended and assumed data.

The failure risk of a pipeline is estimated as the total of all the flaws that are found on it. The Poisson process is used to determine the number of flaws in a pipeline per kilometer. The PCORRC model (Eqs. (8.11) and (8.12)) and the Kiefner model (Eq. (8.13)) are used, respectively, to estimate the rupture pressure and burst pressure in the limit functions.

$$P_{\mathrm{rupture}} = \frac{2\sigma_{\mathrm{f}} W_{\mathrm{t}}}{\mathrm{MD}} \tag{8.11}$$

$$M = \begin{cases} \sqrt{1 + 0.6275\dfrac{L^2}{\mathrm{DW_t}} - 0.003375\left(\dfrac{L^2}{\mathrm{DW_t}}\right)^2} & L \le \sqrt{50W_{\mathrm{t}}} \\ 3.3 + 0.032\dfrac{L^2}{\mathrm{DW_t}} & L > \sqrt{50W_{\mathrm{t}}} \end{cases} \tag{8.12}$$

$$P_{\mathrm{burst}} = \xi \frac{2\sigma_{\mathrm{u}} w_{\mathrm{t}}}{D}\left[1 - \frac{d}{w_{\mathrm{t}}}\left(1 - \exp\left(\frac{-0.157L}{\sqrt{\frac{D(w_{\mathrm{t}}-d)}{2}}}\right)\right)\right] \tag{8.13}$$

in which the signs in the equations are: M—the Folias factor, which explains pipe bulging before collapse; σ_f—the flow stress, (MPa), yield strength-related material property; L—defects length, (mm); D—diameter of pipe, (mm); W_{t}—Wall thickness of pipes, (mm); P_{rupture}—rupture pressure during failure, (MPa); P_{burst}—pressure of failure during burst (MPa); ξ error term for a multiplicative model.

Based on TGNET, a thermal–hydraulic study is used to calculate the internal pressures of pipes. To disperse an operation pressure for every flaw, we randomly choose different space places.

Engineers can evaluate the failure probability of various modes for LNG terminals and empirical evaluation.

For the natural gas pipeline system under discussion, a stochastic capacity network model was created based on the findings of the unit failure test and the capacity estimates. The development process is well illustrated.

Three different scenarios are used to evaluate the supply dependability.

A. Every natural gas supply source is operating normally. This is the standard scenario.
B. The system eliminates LNG terminal A, which had a maximum capacity of 7.1 MCM/d.
C. Removed from the system is LNG terminal B, which has a maximum capacity of 10 MCM/d.
D. The system removes the UGS with a maximum throughput of 4 MCM/d.
E. The system is free of the pipeline that linked nodes 10 and 2.

The states or supply capabilities of the units might fluctuate erratically in various circumstances, as represented by the Markov chains. The Underground Gas Storage (UGS) and LNG terminal are all predicted to experience function degeneration at rates of 0.020, 0.015, and 0.015 (per month), according to the operator of a natural gas pipeline firm. The likelihood of the LNG terminal and UGS experiencing functional interruption is estimated to be 0.001.

For each scenario, one million Markov chains modeling were created, and the state of the supply–demand situation was calculated using the approach created for each one. While the statistics on natural gas flow and demand were given on a regular basis, the modeling was run on a monthly basis.

The three scenarios' cumulative global capacity distribution functions are displayed in Fig. 8.10.

Table 8.1 provides information on the supply capacity for each scenario, including the max, min, mean, and standard deviation.

The study's findings indicate that the pipeline network under consideration has a strong capability to sustain the world's supply at 41.21 MCM/d. Jumps in the CDF are correlated with the impact of random, unforeseen occurrences on the supply capacity of the world. Most "CDF jumps" are small in the probability value range from 10–4 to 10–2, whereas visible "jumps" are seen in the probability value area from 10–6 to 10–4. We can conclude that, even in the event of unanticipated, stochastic unit failures, the gas pipeline network under consideration can dependably provide gas to the consumers.

By comparing the two scenarios, we can see that LNG terminals at various locations play distinct players in the system's supply ability and have various effects on various consumers.

Figure 8.11 provides a study of supply reliability and crisis by combining probability and consequence.

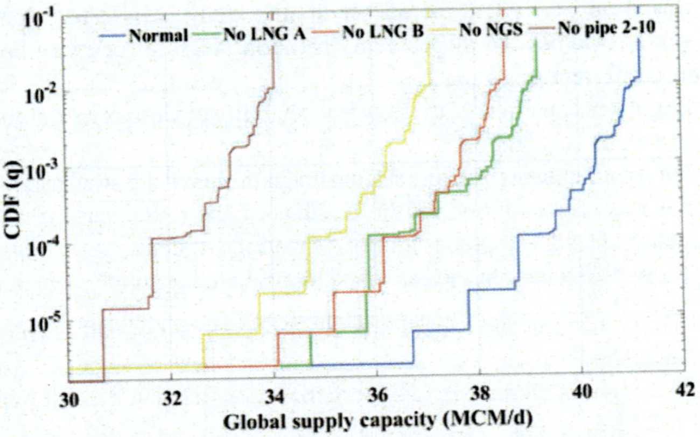

Fig. 8.10 Supply dependability as reflected by the CDF of the world's supply capacity

Table 8.1 Information on the global supply capacity

Scenes	Max (MCM/d)	Min (MCM/d)	Mean (MCM/d)	Std
A	41.21	34.08	41.20	0.061
B	39.21	30.42	39.20	0.072
C	38.58	31.45	38.56	0.061
D	37.13	30.00	37.12	0.059
E	34.08	29.58	34.07	0.056

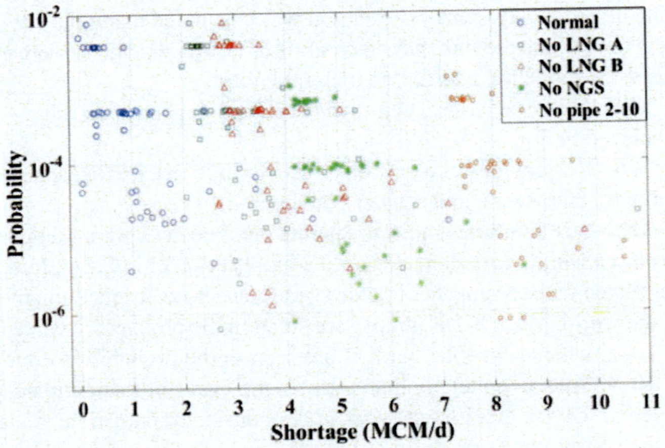

Fig. 8.11 Consequence/probability plot for the supply of gas reliability/risk analysis

Table 8.2 Reliability of the global system supply

Scenario	Average global shortage (MCM/month)	Average amount of unsatisfied customers (per month)
A	0.0049	0.0118
B	0.0365	1.0128
C	0.0479	2.0118
D	0.0743	3.0103
E	0.1285	0.9907

The gas shortage, which has a chance of about 10–6 and would affect about 7.2 MCM/d of gas, would be the worst outcome under normal circumstances. The majority of the effects are centered in the 0–2 MCM/d range, and over half of them have probabilities less than 10–4. The effects of scenes B and C shift toward large amounts of service loss. The average of the global shortfall and the average of dissatisfied customers are computed to determine how much service has been lost. With the results presented in Table 8.2, operators and managers may determine if a system or development is required by measuring the potential loss under the given circumstance. Additionally, one can discover the various effects of various forms of unit failure by contrasting the outcomes of situations A, D, and E. While eliminating a pipeline that has a local, more serious consequence, removing the UGS has a relatively wide-scale. This finding indicates that the pipeline grid is flexible and has strong global resilience, but some customers are exposed due to the vital pipes that connect them to the sources.

However, due to the variations in demand, geography, and other properties across consumers, the ability to service all of them reliably does not necessarily equate to high supply reliability for each individual client (Fig. 8.12). Therefore, in order to obtain a more thorough result, it is necessary to make a specific assessment of each location of need.

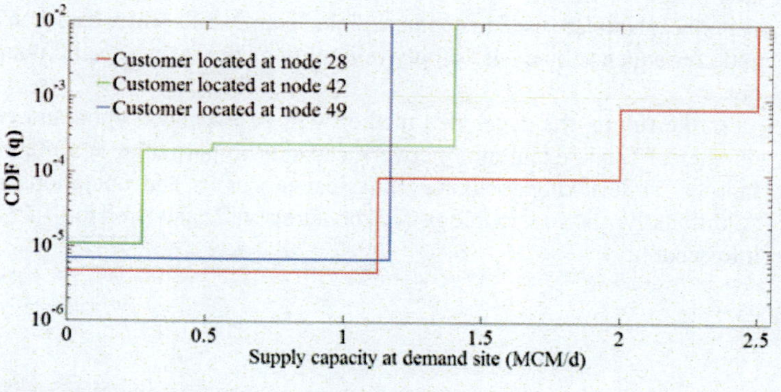

Fig. 8.12 Supply capacity at various demand locations in scenario A

The statistical parameters of increased supply at each demand location are computed using the simulation results and data on each customer's demand.

Two levels of scarcity are used in the analysis: a minor shortage (0.75demand supply capacity requirement) and a severe shortage (0.75demand supply capacity requirement). Every demand location has two degrees of supply dependability, called urgent reliability and normal reliability, which are determined by the rates of severe shortage and moderate shortage, respectively.

Customers' perspectives on dependability provide a more comprehensive and thorough evaluation of the system's capacity to reliably service each client. Along with probability results, quantified losses taking into account the amount of gas and the length of the passive effects are also supplied for each customer, which aids in identifying issues that are challenging to identify from a global perspective, like those discussed in the previous paragraph.

8.4 Conclusions

In order to assess the complex link between various influencing elements and failure modes for unit failure analysis of pipelines, structure reliability analysis methods are applied. The outcomes of an internal examination can be used to gather the data for limited state functions. Failure analysis is carried out based on historical data for compressor stations and natural gas storage equipment.

Based on random processes and graph theory, a stochastic simulation model is built for the investigation of natural gas network capacity and supply consequences. System description, planning, and consequence analysis all employ graph theory as their foundation. The Markov process modeling is used to mimic the system's stochastic evolution through time.

For the supply reliability evaluation, which considers the global system and individual consumers, reliability modeling and principles drawn from the reliability assessment of the power system are employed. The statistical characteristics of system capacity are used to estimate global supply reliability; based on demand data and consequence analysis, supply reliability analysis for specific clients is carried out.

In the future, the developed method will be expanded upon and enhanced in a number of ways, examining network capacity in particular and the effects of unit failures while taking temporary physical processes and operations into account. Additionally, the methodology for modeling will take probabilistic load demand into account.

References

1. O. G. Austvik, "The Energy Union and security-of-gas supply," Energy Policy, vol. 96, pp. 372–382, 2016.
2. E. Zio, An Introduction to the Basics of Reliability and Risk Analysis, vol. 13. World Scientific Publishing Company, 2007.
3. R. Allan and R. Billinton, "Probabilistic assessment of power systems," Proc. IEEE, vol. 88, no. 2, pp. 140–162, Feb. 2000.
4. H.-D. Mo, Y.-F. Li, and E. Zio, "A system-of-systems framework for the reliability analysis of distributed generation systems accounting for the impact of degraded communication networks," Appl. Energy, vol. 183, pp. 805–822, 2016.
5. N. Bassamzadeh and R. Ghanem, "Multiscale stochastic prediction of electricity demand in smart grids using Bayesian networks," Appl. Energy, vol. 193, pp. 369–380, 2017.
6. H. Verdejo, A. Awerkin, E. Saavedra, W. Kliemann, and L. Vargas, "Stochastic modeling to represent wind power generation and demand in electric power system based on real data," Appl. Energy, vol. 173, pp. 283–295, 2016.
7. S. Garshasbi, J. Kurnitski, and Y. Mohammadi, "A hybrid Genetic Algorithm and Monte Carlo simulation approach to predict hourly energy consumption and generation by a cluster of Net Zero Energy Buildings," Appl. Energy, vol. 179, pp. 626–637, 2016.
8. C. Hu, S.-Y. Lu, and Y.-H. Chen, "Stochastic–multiobjective market equilibrium analysis of a demand response program in energy market under uncertainty," Appl. Energy, vol. 182, pp. 500–506, 2016.
9. M. Marseguerra and E. Zio, "Monte Carlo approach to PSA for dynamic process systems," Reliab. Eng. Syst. Saf., vol. 52, no. 3, pp. 227–241, 1996.
10. X. Fu, H. Sun, Q. Guo, Z. Pan, X. Zhang, and S. Zeng, "Probabilistic power flow analysis considering the dependence between power and heat," Appl. Energy, vol. 191, pp. 582–592, 2017.
11. A. Szikszai and F. Monforti, "GEMFLOW: A time dependent model to assess responses to natural gas supply crises," Energy Policy, vol. 39, no. 9, pp. 5129–5136, 2011.
12. M. Flouri, C. Karakosta, C. Kladouchou, and J. Psarras, "How does a natural gas supply interruption affect the EU gas security? A Monte Carlo simulation," Renew. Sustain. Energy Rev., vol. 44, pp. 785–796, 2015.
13. E. G. P. I. D. G. EGIG, "9th Report of the European Gas Pipeline Incident Data Group (1970–2013)," no. 1, p. 23, 2015.
14. S. Zhang and W. Zhou, "System reliability of corroding pipelines considering stochastic process-based models for defect growth and internal pressure," Int. J. Press. Vessel. Pip., vol. 111, pp. 120–130, 2013.

Chapter 9
Risk Prewarning Method for Pipeline Systems

Zhaoming Yang and Lixun Chi

Abstract Oil and gas pipeline networks have complicated, irregular operational factors. Traditional risk prewarning techniques, including time–frequency domain transformation, are good at forecasting changes in data points at a certain instant in time, but they are unable to detect and identify the system's operational condition over an extended period of time. The operational status of an oil and gas pipeline may be accurately identified using a data-driven strategy for identifying time series patterns that is introduced in this chapter. The monitoring procedure and identification outcomes of the pipeline operation may be managed and understood in the meanwhile. The goal of the suggested solution is to offer a comprehensible anomaly warning system for the operation management of product oil pipelines.

9.1 Risk Prewarning of Oil and Gas Pipelines

The conventional approach to evaluating system reliability can be based on the dependability of the system's components and supplemented with models from conceptual network theory. It is challenging for them to use abstract topological structures to represent dynamic and complicated systems.

Some stable physical models are described by state-space equations. Zeng et al. [1] proposed a stable approach with bi-directional energy transfer to analyze the energy flow in IESs; however, they overlooked the optimization of various energy systems. The energy flow optimization was then taken into consideration in a condensed version by Wang et al. [2] using a linearized modeling approach. When modeling the IES, linear programming used by Qadrdan et al. [3] is to solve non-linear problems.

Z. Yang (✉)
National Engineering Laboratory for Pipeline Safety; Beijing Key Laboratory of Urban Oiland Gas Distribution Technology, China University of Petroleum, 102249 Beijing, China
e-mail: yangzhaoming.upc@foxmail.com

L. Chi
PetroChina Planning & Engineering Institute, 100083 Beijing, China
e-mail: chilixun@icloud.com

H. Su et al. (eds.), *Advanced Intelligent Pipeline Management Technology*,
https://doi.org/10.1007/978-981-19-9899-7_9

The bi-direction energy flow was disregarded despite the adoption of the IES data-driven model [4]. For this reason, an effective data-driven approach is suggested in this chapter that can precisely describe the behavior of complicated bi-directional IESs.

In order to employ realizations of samples with unknown variables in the reliability evaluation in view of these distributional functions of probability, one may use the Monte Carlo method. To describe the inherent unpredictability of the power demand, Zio et al. [5] employed a probabilistic load model. In the literature [6], uncertain heat loads are represented by a normal distribution, but the gas transport capacity is represented by a lognormal distribution. Su et al. [7] use the Markov process to assess the dependability of the IES's energy deliver. To address this problem, Ciupăgeanu et al. [8] limited their assessment of the impact of integrating substantial percentages of renewable energy on system performance and dependability to standard scenarios.

Data in time series are therefore required for real-time dependability evaluations. Clegg et al. [9] studied the impact of electricity to natural gas conversion on operating system reliability using historical daily demand and forecast renewable energy production. In order to increase the dependability of the energy supply in IES, Alabdulwahab et al. [10] examined the cooperation between the restricted electrical and buildings for storing natural gas.

To explain the uncertainties in variables, probabilistic forecasting techniques can provide densities of probabilities, prediction intervals (PIs), and quantiles. Pinson and Kariniotakis [11] created a method for calculating the intervals for the predictions (PIs) of wind power generation that is founded on fuzzy inference and adaptive resampling.

Diagnostic and degradation analysis techniques for systems and components are based on process variables, such as those found in rotating equipment and wind systems. For complex system dependability evaluations based on health status indicators, dynamic Bayesian networks are frequently utilized. It is difficult to assess the functioning status of the systems based on several process factors. Combined condition indicators, the hidden Markov Model (HMM) may calculate the system probability distribution and take time-dependent internal relations between variables into account.

With the help of the suggested strategy, the uncertainty issue may be resolved by animatedly predicting the mathematical distribution of important parameters over time. On the basis of the process variables, HMM is offered to analyze the functional states of the system. It is possible to describe relationships between states and variables using the formats of hidden state and transition probabilities.

9.2 Methodology

In order to assess the dependability of the energy supply in IES, a systematic framework is created. Data predicting, IES modeling, modeling of system functional states, and supply dependability evaluation are the four steps of the technique, as illustrated in Fig. 9.1.

Step 1. In order to predict the renewable energy output PI, a probabilistic forecasting model was utilized. This model is based on the ELM and the pairwise bootstrap method.
Step 2. A dynamic simulation model based on stacked autoencoders is used to estimate how well the IES will perform dynamically. As inputs, the probabilistic forecasting model from the earlier step uses samples of the prediction data.
Step 3. Using the outputs from the dynamic simulation model from the step before, such as delivery pressure at the customer nodes, total gas requirements, using HMM to predict the operational state of a system.
Step 4. Continue using Step 2 until Order Statistics has got the necessary percentiles for each state's probability distribution.

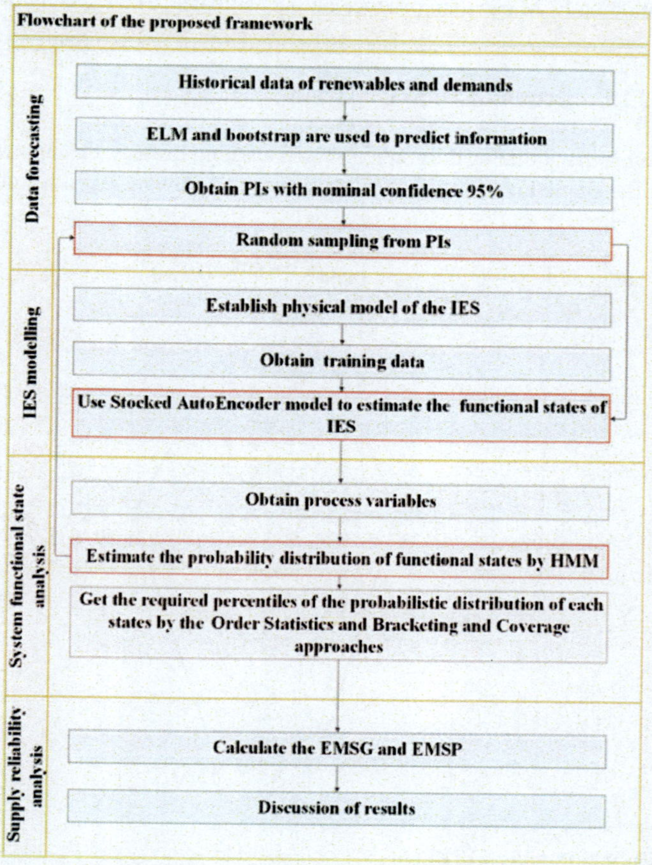

Fig. 9.1 Flowchart of the proposed framework

Step 5. Finally, the outcomes of the system functional states are used to determine the energy supply's dependability.

9.2.1 *Probabilistic Forecasting Model*

For a system to operate effectively and dependably, projecting renewable energy output and demand must be accurate and dependable. The LASSO-QRNN model, the aggregated probabilistic forecasting approach, the Gaussian process quantile regression model, and the exponential smoothing method are just a few of the probabilistic forecasting methods that have been created. Faster than previous learning algorithm-based forecasting techniques, the bootstrap-based extreme learning machine (BELM) can estimate the PIs. Figure 9.2 depicts the BELM architecture for constructing PIs. By resampling from the data, the paired bootstrap is utilized to acquire the practice data. The ELM can then produce output estimates. The PIs can now be predicted.

9.2.1.1 Extreme Learning Machine

Rather than utilizing the gradient-based learning strategy, the input weight and biases of the single hidden layer feedforward neural network ELM are chosen at random.

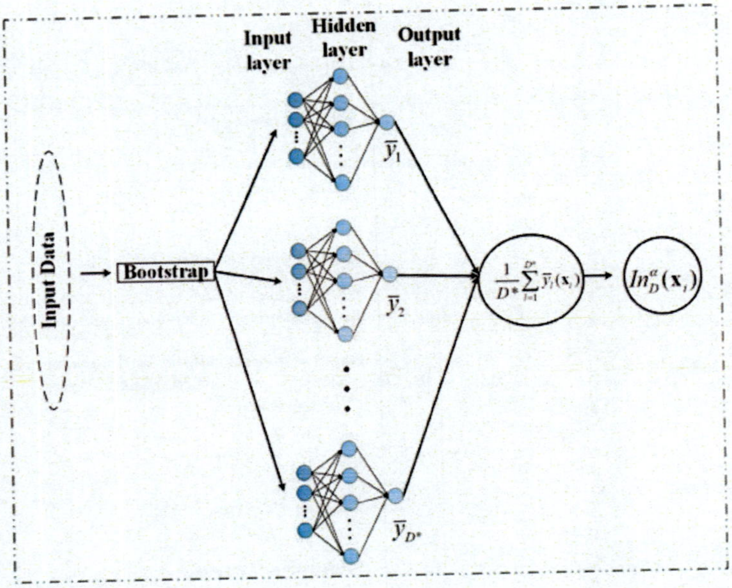

Fig. 9.2 Framework of BELM

The output weight may be calculated using a straightforward matrix formula. The following is an explanation of the ELM training steps:

(1) Generated the hidden layer nodes and input weights randomly.
(2) Based on the hidden layer's neuron activation function, determine the output matrix H.
(3) The weights between the output layer and the concealed layer should be calculated.

9.2.1.2 Prediction Interval

The ELM may be used to identify the connections between the inputs and outputs, after which the pairs of bootstrap technique are used to estimate the PIs. Model underspecification and noise in the data are the two basic categories of defects that may be taken into account. The average value of the guesses may be taken into account as the actual regression, given the bootstrapped pairings D* based on the initial training data.

9.2.2 Dynamic Model of IES

The model's central concept is to recreate the input at the autoencoder's output, as seen in Fig. 9.3. Encoding and decoding are the two pieces that carry out this procedure. The encoder collects the features of the data and maps the input to the hidden layer. The collected features are remapped to the original input space by the decoder during a reconstruction process.

Encoder and decoder operations can be shown as:

$$z = E(\boldsymbol{x}, \boldsymbol{\theta}) \tag{9.1}$$

$$\boldsymbol{x}' = D(z, \boldsymbol{\theta}') \tag{9.2}$$

the $\boldsymbol{x} \in \boldsymbol{R}^n$ and $\boldsymbol{x}' \in \boldsymbol{R}^n$ represent the inputs and the resulting redevelopment; z represents the indirect representation E and D represent the activation works in accordance with the parameter $\boldsymbol{\theta}$ and $\boldsymbol{\theta}'$, respectively, including deviation vector and weight matrix. The mathematical expression of the loss function, which is used to rebuild the raw inputs from the compressed characteristics, is as follows:

$$L(\boldsymbol{x}, \boldsymbol{x}') = \|\boldsymbol{x} - \boldsymbol{x}'\|^2 \tag{9.3}$$

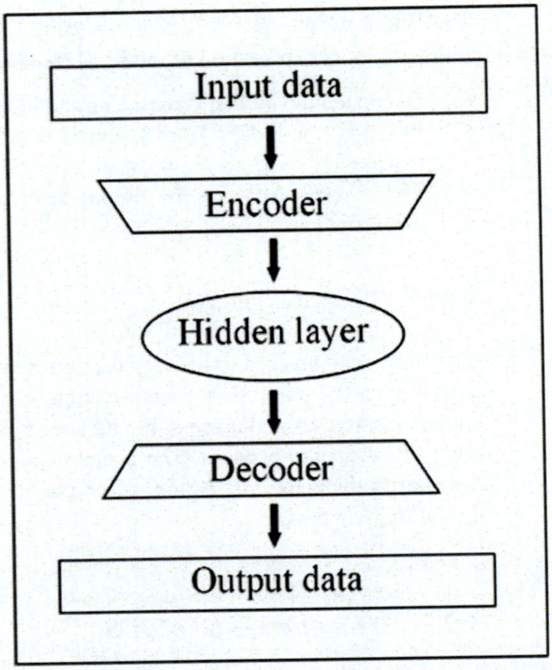

Fig. 9.3 Basic autoencoder model's structure

9.2.3 Modeling of System Functional States

9.2.3.1 HMM

While accounting for a number of IESs features, HMM is used to analyze the system functional states and determine the likelihood function that occupies every functional state. The availability of the system is determined by the likelihood of the system's functioning states.

Make a functional state-space definition $F(D_1, D_2, D_3, \ldots, D_n)$, , where the Gaussian mixture model (GMM) is followed by the functional indicator vector D_i.

Equation (9.4) may be used to analyze the stochastic properties of the time series data and model the combined variable probability distribution:

$$p(\boldsymbol{OZ}|\lambda) = p(z_1|\boldsymbol{\pi})\prod_{s=2}^{S} p(z_s|z_{s-1}, \boldsymbol{A})\prod_{s=1}^{S} p(\boldsymbol{o}_s|z_s, \phi) \tag{9.4}$$

where $\boldsymbol{o} = \{o_1, o_2, o_3, \ldots, o_E\}$ and $z = \{z_1, z_2, z_3, \ldots, z_H\}$ represent the observable and latent vector; H is how many states are there in a Markov chain; E is the number of observational variables that may be present in each state; $\boldsymbol{\pi}$ is the primary probability factor; $\boldsymbol{A}$ is the probabilities matrix for state transitions, whose components are the chances that two hidden states will change; ϕ is the parameter associated with the observation distribution; the compact symbol $\lambda = \{\boldsymbol{\pi}, \boldsymbol{A}, \phi\}$ is used to describe the set of parameters. ϕ can be defined as $\phi = \{B_{hm}, \mu_{hm}, \boldsymbol{\Sigma}_{hm}\}$, where B_{hm} is the probability of the mth mixture under the hth state; $\boldsymbol{\mu}_{hm}$ and $\boldsymbol{\Sigma}_{hm}$ are the mean vector and covariance vector of the mth mixture in the hth state, respectively. Based on the cumulative distribution function, ϕ can be calculated by Eq. (9.5):

$$p(o|z_h\phi) = \sum_{m=1}^{M} B_{hm} N\left(\boldsymbol{o}\middle|\boldsymbol{\mu}_{hm}, \boldsymbol{\Sigma}_{hm}\right) \tag{9.5}$$

The expectation maximization (EM) technique may estimate the parameter, mean vector, and covariance vector based on historical data.

The state occupancy probabilities $\gamma(z_{sh})$ and $\varepsilon(z_{s-1,j}z_{s,h})$ are iteratively determined in the expectation stage. $\gamma(z_{sh})$ can be decomposed into a forward term $\hat{\alpha}(z_s)$ and a backward term $\hat{\beta}(z_s)$ by Eq. 9.6:

$$\begin{cases} \hat{\alpha}(z_s) = \dfrac{1}{c_s} p(\boldsymbol{c}_s|z_s) \displaystyle\sum_{\boldsymbol{z}_{s-1}} \hat{\alpha}(z_{s-1}) p(z_s|z_{s-1}) \\ \hat{\beta}(z_s) = \dfrac{1}{c_{s+1}} \displaystyle\sum_{z_{s+1}} p(\boldsymbol{c}_{s+1}|z_{s+1}) \beta(z_{s+1}) p(z_{s+1}|z_s) \\ \gamma(z_s) = \hat{\alpha}(z_s)\hat{\beta}(z_s) \\ \varepsilon(z_s) = \hat{\alpha}(z_s)\hat{\beta}(z_s) p(\boldsymbol{c}_s|z_s) p(\boldsymbol{c}_s|z_{s-1}) \\ \quad c_s = p(z_t|z_1, z_2, \ldots, z_{s-1}) \end{cases} \tag{9.6}$$

Another latent variable, $\boldsymbol{v}_t$, dominates the probability distribution of the observation symbol in GMM-HMM. $\eta(\boldsymbol{v}_s z_s)$ can be obtained by Eq. 9.7:

$$\begin{cases} \eta(\boldsymbol{v}_s z_s) = \hat{\alpha}(\boldsymbol{v}_s z_s)\hat{\beta}(\boldsymbol{v}_s z_s) \\ \hat{\alpha}(\boldsymbol{v}_s z_s) = \dfrac{1}{c_s} p(\boldsymbol{o}_s|\boldsymbol{v}_s) p(\boldsymbol{v}_s|z_s) \displaystyle\sum_{\boldsymbol{v}_{s-1} z_{s-1}} \hat{\alpha}(\boldsymbol{v}_{s-1} z_{s-1}) p(z_s|z_{s-1}) \\ \hat{\beta}(\boldsymbol{v}_s z_s) = \dfrac{1}{c_{s+1}} \displaystyle\sum_{\boldsymbol{v}_{s+1} z_{s+1}} \hat{\beta}(\boldsymbol{v}_{s+1} z_{s+1}) p(\boldsymbol{o}_{s+1}|\boldsymbol{v}_{s+1}) p(\boldsymbol{v}_{s+1}|z_{s+1}) p(z_{s+1}|z_s) \end{cases} \tag{9.7}$$

For maximization, Eq. 9.8 re-estimates the parameters:

$$\begin{cases} \pi_h = \dfrac{\gamma_{1h}}{\sum\limits_{j=1}^{H}\gamma_{1j}}, \quad k = 1, 2, \ldots, H \\ A_{jh} = \dfrac{\sum\limits_{t=2}^{T}\varepsilon(z_{t-1,j}z_{t,h})}{\sum\limits_{l=1}^{H}\sum\limits_{t=2}^{T}\varepsilon(z_{t-1,h}z_{t,l})}, \quad j, h = 1, 2, \ldots, H \\ B_{hm} = \dfrac{\sum\limits_{s=1}^{S}\eta_{shm}}{\sum\limits_{s=1}^{S}\sum\limits_{j=1}^{M}\eta_{shj}}, \boldsymbol{\mu}_{hm} = \dfrac{\sum\limits_{s=1}^{S}\eta_{shm}\boldsymbol{o}_s}{\sum\limits_{s=1}^{S}\eta_{shm}}, \boldsymbol{\Sigma}_{hm} = \dfrac{\sum\limits_{s=1}^{S}\eta_{shm}(\boldsymbol{o}_s - \boldsymbol{\mu}_{hm})(\boldsymbol{o}_s - \boldsymbol{\mu}_{hm})^T}{\sum\limits_{s=1}^{S}\eta_{shm}} \end{cases} \tag{9.8}$$

According to the functional indicators, it is feasible to calculate the distribution of occupancy probabilities for every system functional state. The likelihood of being in a particular condition at a particular moment can be thought of as the forward term $\hat{\alpha}(z_s)$. After that, using the variables of the continuous time process, it is feasible to calculate the system's properly functioning states' probability distribution at a certain time.

9.2.3.2 Order Statistics

Here, Order Statistics (OS) is employed to lessen the necessary computing load. The result y should indicate the likelihood that functional state h will exist at time t_0. For various input vectors $X = \{x_1, x_2, x_3, \ldots, x_Q\}$, the corresponding output vectors $Y = \{y_1, y_2, \ldots, y_Q\}$ may be produced with Q runs of the GMM-HMM. By using the OS technique, an envelope of the likelihood of occupying a functional state at time t_0 may be defined at a certain degree of confidence. The output vectors $Y = \{y_1, y_2, \ldots, y_Q\}$ are organized as $Y^* = \{y^1, y^2, \ldots, y^q\}$ in decreasing value order. The qth member y_q has a chance of ζ being higher than the unidentified real γth percentile y_γ. The actual value of y_γ provides some assurance that the value will be less than y_q. This means that the real y_γ will likewise be lower if the value of y_q is less than a certain upper threshold. Additionally, y_γ may be above the lower barrier as shown by Eq. (9.9):

$$\begin{cases} \gamma_1 = P\{y < y_{\gamma_1}\} \\ \zeta_1 = P\{y_{\gamma_1} < \hat{y}_{\gamma_1}\} \end{cases}, \quad \begin{cases} \gamma_2 = P\{y < y_{\gamma_2}\} \\ \zeta_2 = P\{y_{\gamma_1} > \hat{y}_{\gamma_2}\} \end{cases} \tag{9.9}$$

where $\hat{y}_\gamma$ is the γth percentile.

The Bracketing technique may be used to determine the required sample size Q once the confidence level has been established. Following is a description of the steps:

Step 1: Determine γ_1, ζ_1 and Q.
Step 2: The output Y may be approximated using the vectors X sampled from the relevant distribution.
Step 3: The inputs will be regenerated each time Step 1 is repeated G times. The outcomes can be found as $\hat{y}_q^g$ (g = 1, 2, …, G, n = 1, 2, …, Q).
Step 4: The (ζ|γ) approximate percentile $\hat{y}_\gamma^g$ of each output sample can be computed.
Step 5: The median ${}_{0.5}\hat{y}_\gamma$ of the G(ζ|γ) percentile ranges $\hat{Y} = \{\hat{y}_\gamma^1, \hat{y}_\gamma^2, \ldots, \hat{y}_\gamma^G\}$ has been applied to indicate the guessed amount.

9.2.4 Supply Reliability Analysis

Two processes make up the supply reliability evaluation process: estimating the functional state probability distribution and computing the system reliability index. HMM and OS complete the first stage, as was already described. The dependability of IES is then examined in the following phase using a combination of an adequate indication and a security indicator.

The adequate indicator shows how successfully IES manages to maintain the supply–demand connection within the predetermined tolerances. Equation (9.10) suggests the increased indicators (9.10):

$$\text{SAIFI} = \frac{\sum_i \lambda_i^f \text{CN}_i}{\sum_i \text{CN}_i} \tag{9.10}$$

where CN_i is quantity of loads; λ_i^f is the client node's failure rate. The estimated max shortages of gas and power, respectively, are denoted by the abbreviations EMSG_t and EMSP_t.

$$\begin{aligned}&\text{EMSP}_t/\text{EMSG}_t\\&=\sum_{h=1}^{H}\int_{Var_{t,l,i}}^{Var_{t,u,i}}\int_{P_{t,l,h}}^{P_{t,u,h}} As_h\frac{Var_{t,i}\times p_{t,h}}{(P_{t,u,h}-P_{t,l,h})\times(Var_{t,u,i}-Var_{t,l,i})}\mathrm{d}p_{t,h}\mathrm{d}Var_{t,i}\end{aligned} \tag{9.11}$$

where $P_{t,l,h}$ and $P_{t,u,h}$ are the probability interval's lower and upper bounds in the state h. $P_{t,h}$ is the likelihood at time t. $Var_{t,i}$ is what the associated variables were worth at time t. $Var_{t,j,i}$ and $Var_{t,u,i}$ are the linked variables' lowest value and maximum value, respectively. A_h is the risk coefficient.

The dependability index, CDI, measures an IES's capacity to function as needed even in the event of an accident within a defined period of time. Equation (9.12) provides a description of the equation:

$$\begin{cases} RS(W)_{h,s,t} = \dfrac{MV(\text{period}) - \text{WorstMV}}{\text{BestMV} - \text{WorstMV}} \\ RS(W)_{h,t} = \sum\limits_{s=1}^{S} Wt(s)RS(W)_{h,s,t} \\ RS(W)_{t} = \sum\limits_{h=1}^{H} Wt(h)RS(W)_{h,t} \end{cases} \tag{9.12}$$

in which RS stands for relative score; W reflects a variety of variables, including flow rate, S is a sort of index that contains the probability distribution's upper and lower bounds, interval, and range; Wt is the weight; MV is the index's time value s; WorstMV and BestMV are the index's worst values and highest values s, respectively, the results of which are derived from the past data. A weighted average of the relative scores may be used to determine CDI over a specific time period. Equation 9.13 is as follows:

$$CDI_t = \sum Wt(W) \times RS(W)_t \tag{9.13}$$

where $Wt(W)$ is a weight of W, which can be calculated by the principle [12].

9.3 Case Study

9.3.1 Data Description

In this chapter, ENTSOE is used to retrieve historical data on electricity usage and renewable energy production. The statistics on gas consumption are compiled from figure sharing, an open-access public database.

9.3.2 Description of the IES Model

The efficacy of the suggested approach is assessed using a proven bi-directional IES model. Figure 9.4 displays the IES model, which consists of an IEEE-12 test system and a 12-node natural gas network. Node N1 serves as a gas source for the majority of the gas. There are wind farms at Bus B7 and Bus B10, solar power generators at Bus B15, gas-fired generators at Bus B9 and Bus B13, a coal-fired power plant at

Bus B4, and other electrical infrastructures that are built to accommodate the power needs. Additionally, three separate pieces of machinery—gas-fired generators, gas compressors, and P2G—joined the electrical grid with the gas system.

By simulating this model, it is possible to acquire process parameters such as voltage, fluid velocity, energy losses, and pressure difference. Table 9.1 presents the four groups into which they are divided.

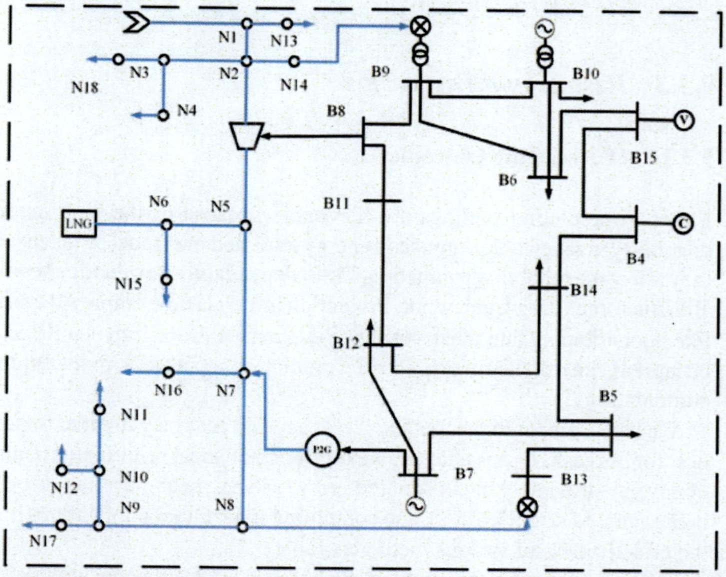

Fig. 9.4 IES model's architecture

Table 9.1 Process variables connected to four groups

Group	Process parameters (functional indicators)
A	Rate of gas source flow m^3/s
B	Pressure of nodal gas kPa
C	Loss of transmission kW
D	P2G energy conversion m^3/s GPG energy conversion kW

Table 9.2 Outcomes of various methods for making predictions

Methods	MAE	MSE	RMSE	Time (s)
ELM	16.4389	378.1296	19.4610	0.3055
LSTM	17.4752	460.5956	21.4615	25.1532
ANN	11.5002	233.6488	15.2856	0.5201
Autoregressive moving average model (AMRA)	45.7174	2672.8956	51.7001	81.7285
Backpropagation neural network (BP)	44.3163	2700.3002	51.9644	2.6521

9.3.3 *Results and Discussion*

9.3.3.1 Forecasting Outcomes

Errors in predicting findings are inevitable because of the high degree of unpredictability in renewable output and energy load, and they must be taken into account as they affect the reliability evaluation. Thus, dependability evaluation requires the probabilistic forecasting framework. Prediction intervals now employ Bayesian, methods like bootstrapping and mean–variance estimation. Bootstrap is utilized in this study because it can efficiently estimate the prediction variance without requiring intricate computations.

We have compared the methods' efficacy and accuracy in order to decide which is best for our task. In this chapter, we use four statistical criteria to examine the results of several strategies: mean squared error (MSE), mean absolute error (MAE), root mean squared error (RMSE), and computing time. Other publications have compared the effectiveness of various methods (Table 9.2).

Table 9.2 demonstrates that while ELM's accuracy may not always be the greatest and most effective, it does need the least amount of calculating time (0.3055 s). Instead, the ANN method produces greater prediction results. On the flip side, because the bootstrap method's calculation of PIs takes a lengthy time, computation time is an important factor. The BELM approach is utilized for probabilistic forecasting, since the PIs using the bootstrap methodology in any case of the uncertainty linked to the method accuracy. The PIs of renewable energy generation and energy consumption can be used to evaluate supply dependability. The selected confidence level in this instance is 95%. By taking a sample from the connected PIs, which serve as the simulation model's inputs, one may determine the point value of every variable.

9.3.3.2 Results of IES Dynamic Performance Simulation

Since labeled data collection is a realistic challenge and its associated sample imbalance problems complicate model training, unsupervised approaches have been developed in several disciplines. For a variety of tasks, including classification, behavior analysis, machine translation, etc., convolutional neural networks (CNNs), recurrent

Table 9.3 Comparison of prediction results of different methods

Methods	MAE	MRE	RMSE
SAE	0.0894	0.3997	0.0046
CNN	9.3002	0.4535	14.4368
RNN	45.4078	1.6002	71.51.2

Table 9.4 Calculation time using different methods

Methods	Time (s)
SAE	2.8640×10^{-5}
CNN	3.4145×10^{-4}
RNN	9.7560×10^{-5}
Physical model	2.9264

neural networks (RNNs), deep belief networks (DBNs), LSTM, and stacked autoencoder (SAE) have all been used. The findings produced by DBN, RNN, and CNN are compared with SAE in order to show the adaptation of SAE to IES performance predictions. The physical model of the IES is used to simulate the creation of 2000 sets of process variables, which are then utilized as input samples for all models. Comparing the outcomes of various approaches is provided in Table 9.3.

Table 9.3 demonstrates the effectiveness of SAE in extracting useful feature representations from IES data. Additionally, SAE uses the least amount of computational time (Table 9.4; 0.000977% of the time required for the physical model). Consequently, an SAE model rather than the conventional physical model is employed to boost computing efficiency.

Figure 9.5 and Table 9.5 provide a detailed description of the stacked autoencoder model's dependability and accuracy.

Figure 9.5 demonstrates how the predicting outcomes may precisely mirror the system's dynamic performance. These findings point to a fair degree of accuracy in predicting system dynamic performance.

9.3.3.3 Results of the Estimate of System Functional States

The parameters of the GMM-HMM may be calculated from the functional indicators that have previously been observed. On the basis of continuous time process, hidden states and related probability distributions can be generated and used to investigate the way in which IES operates. Reliability is assessed using estimations of the probabilities of the states. The probability distributions of every state at specified periods may be calculated by combining the Order Statics and Bracketing techniques.

The input for this technique is specified as the variables' means of the PIs. The outcomes are displayed in Fig. 9.6a. The likelihood of failure is expressed as a point value and is, for example, 0.036 at hour 8. But it is insufficient for operators to make sound decisions. Due to the uncertainty elements that have been taken into account,

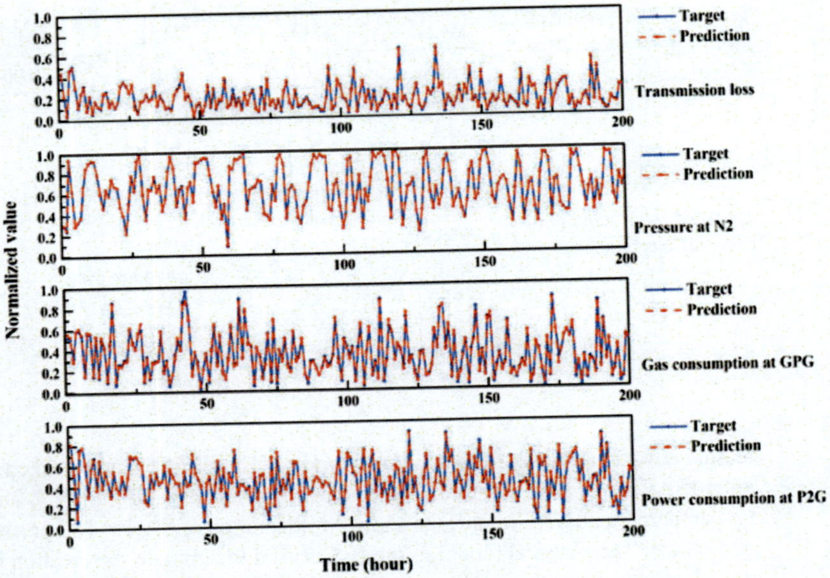

Fig. 9.5 Forecasting results of different variables

Table 9.5 Outcomes of the prediction using the stacked autoencoder model

Process variables	MAE	MRE	RMSE
Delivery pressure	0.0026	0.0052	0.0002
Transmission loss	0.0070	0.0621	0.0007
Consumption of GPG	0.0163	0.0870	0.0015
Consumption of P2G	0.0009	0.0421	0.0010

the hourly probability distributions of each condition in different groups at hour 6 (as shown in Fig. 9.7) can offer more accurate information.

Additionally, as can be shown in Fig. 9.6a, the typical dependability evaluation only offers a number that captures the system's current condition. The findings are different as indicated in Fig. 9.6b since the input is varied, demonstrating that using only a time series data may not provide very impressive results.

On the other hand, the suggested model allows for the grouping of variables into groups in order to determine the system's state, depending on the priority of experts and operators (as shown in Fig. 9.7).

The probability distributions for each circumstance at any hour may be presented, as seen in Fig. 9.8, which can give more detailed and accurate information. The results are more indicative than those from other approaches.

Figure 9.8 displays the probability distribution time series for each system functional state. Group A's system functionality is mostly dominated by low-risk and medium-risk states. It suggests that if the change in gas supply is taken into account

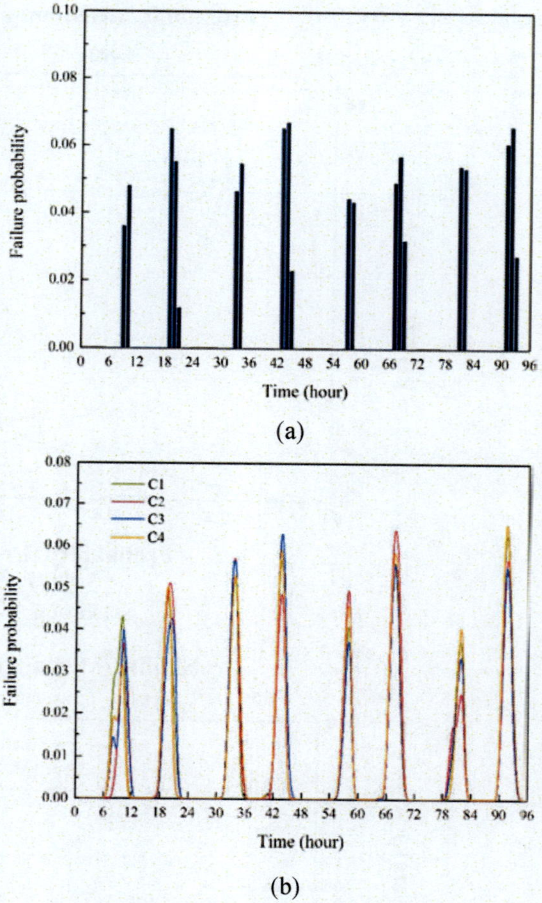

Fig. 9.6 IES failure likelihood determined via dynamic event tree analysis

simply, the system may be kept in a reasonably healthy state. The distribution of low-risk probabilities in Group B varies from 0.98 to 1 during hours 1–4. Other times, the high risk and moderate risk probability distribution ranges are nearly identical. The cause is that after hour 6, consumers begin using natural gas regularly. The transmission loss in Group C, which is affected by both the customer demand and the delivery pressure, is associated with power transmission loss and compressor power usage. Each state's probability distributions in Group D are comparable. This suggests that each state has a distinct tendency. In general, stochastic renewable generation must be modulated by frequent operation of GPG and P2G. Compared to other groups, the functional states are more delicate to the equilibrium between renewable energy production and energy consumption.

Because a set of time series is employed in conventional reliability evaluation methods, only a single value is supplied (as shown in Fig. 9.6). Because projected data

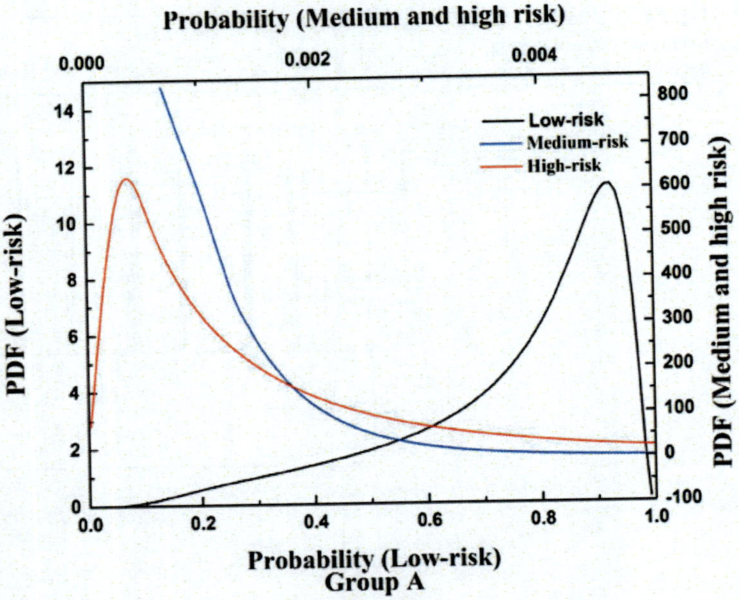

(a) Group A

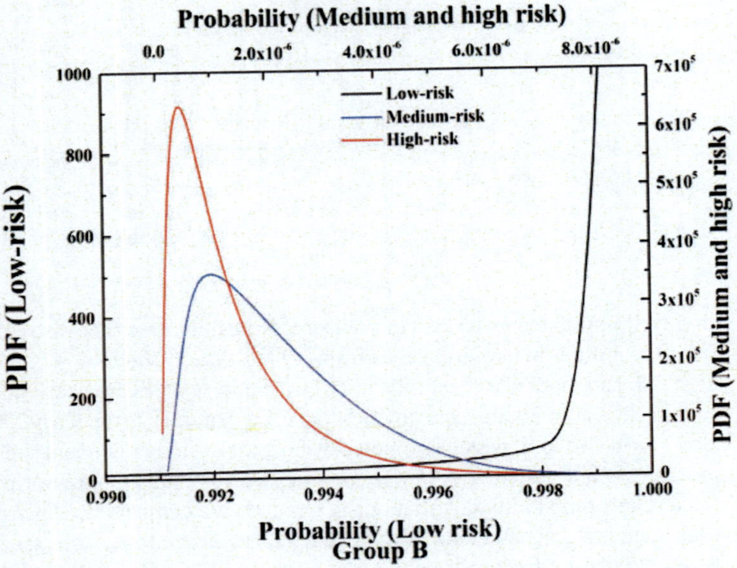

(b) Group B

Fig. 9.7 Probability distribution functions at hour 6

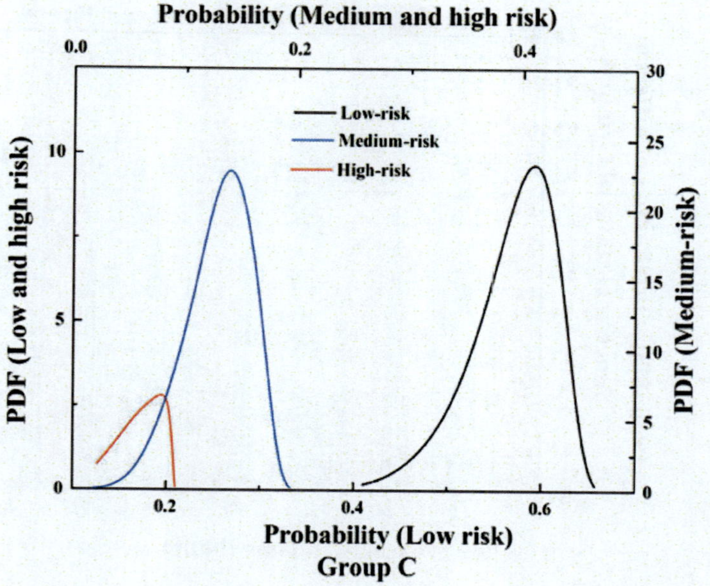

(c) Group C

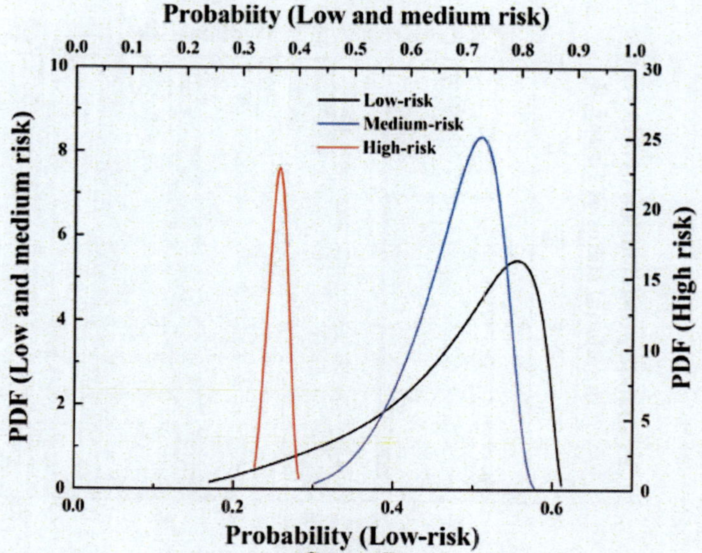

(d) Group D

Fig. 9.7 (continued)

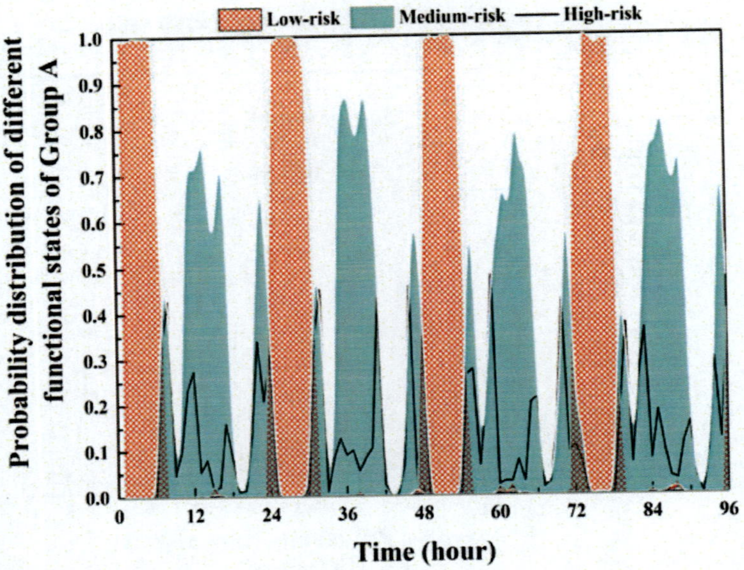

(a) Group A

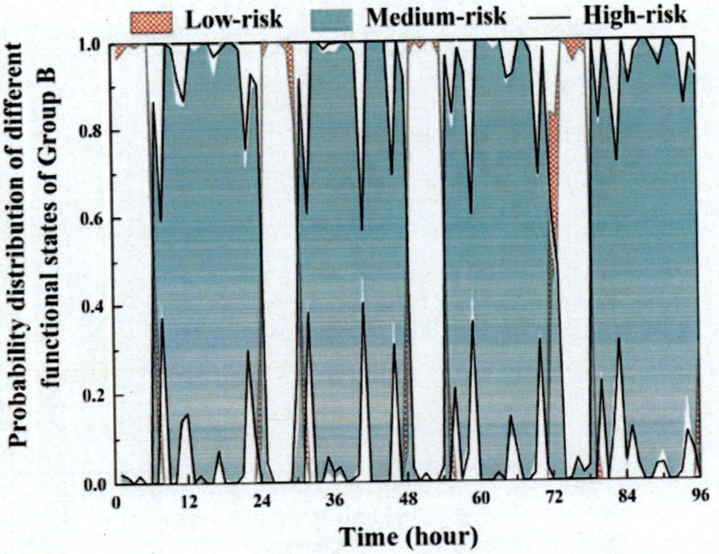

(b) Group B

Fig. 9.8 Probability distributions for each state, with a 95% confidence level

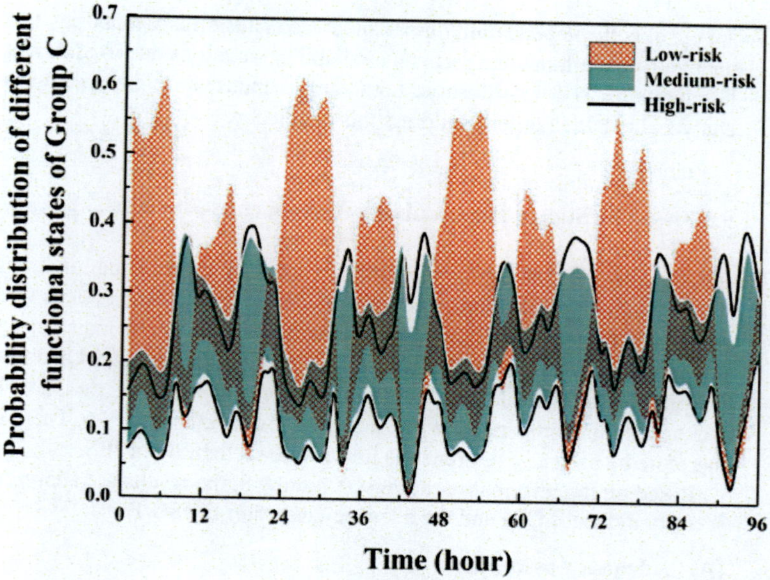

(c) Group C

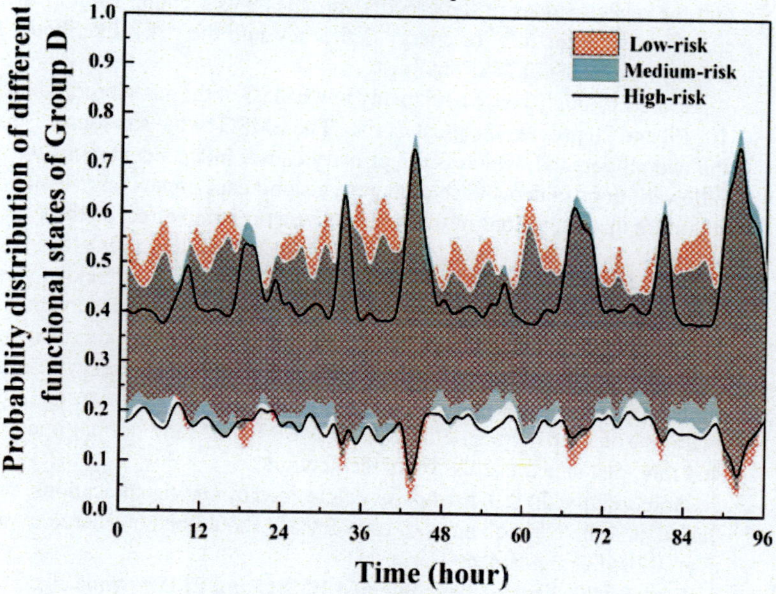

(d) Group D

Fig. 9.8 (continued)

are chaotic, the dependability assessment's conclusion is astounding. Operators are unable to detect flaws and recognize reliability situations due to PIs' poor visibility. Equations (9.10–9.13) illustrate how various indexes are created to combine and analyze the data in order to address this deficiency.

9.3.3.4 The Supply Dependability Assessment's Findings

By comparing and analyzing the specific outcomes, the information in Figs. 9.7 and 9.8 can show how each process variable affects the way the system works. In this research, two indexes (CDI and EMSG/P) are presented to give managers quantifiable results, on the behavior of the system in place of entirely normal and failure states. Despite the fact that these indices are point values, uncertainties have been taken into account in comparison to the outcomes displayed in Fig. 9.6. These indices can therefore be used as a reference to improve the reliability of IES.

Based on the information in Figs. 9.7 and 9.8, the results are analyzed using the security indicator CDI and the adequacy indicator EMSG/P.

(A) **Adequacy assessment**

When compared to conventional reliability indices, EMSG and EMSP take the PIs and the related values of the process variables into account. Both of them are able to translate the quantity of energy deficiency into the risk's likelihood. Figure 9.9 displays the EMSG/EMSP findings.

Hours 1 through 5 have a relatively low EMSG of the gas source, whereas hours 9, 10, 19, and 20 provide the greatest risk. The EMSG of the gas source is impacted by the overall demand, which is the primary cause. Independent consumers' sporadic little variations in demand seldom have a significant impact on overall gas demand. It implies that the actions of the people have a profound effect on the system dependability. It should be observed that the EMSG of GPG fluctuates often, and that daily risk times are more variable than other times. GPG must react quickly to the stochastic renewable's outputs, which frequently cause it to start up and stop down.

The EMSP for transmission loss follows a pattern that is almost identical to the EMSPs for P2G and GPG. Because of the rise in renewable energy generation, it has been reported that the danger hours are more likely to occur around hour 22. The main reasons for frequent changes in EMSP include the normal operation of P2G and long-distance transmission in the network.

Additionally, the risk periods, as determined by various indications, are extremely varied, and the consequences of each EMSG should be considered by the system's overall EMSG and EMSP.

Figure 9.10 displays the combined EMSG and EMSP, while Fig. 9.11 displays the average percentages for each component. The percentages of EMSP (P2G) and EMSP do not fluctuate much over time (approximately 85% and 15%, respectively), despite variations in the combined EMSP's value. The proportions of each specific EMSG, however, can alter drastically. Additionally, at some hours (almost 0% at hour 6), the effect of the gas source on the EMGS can be disregarded.

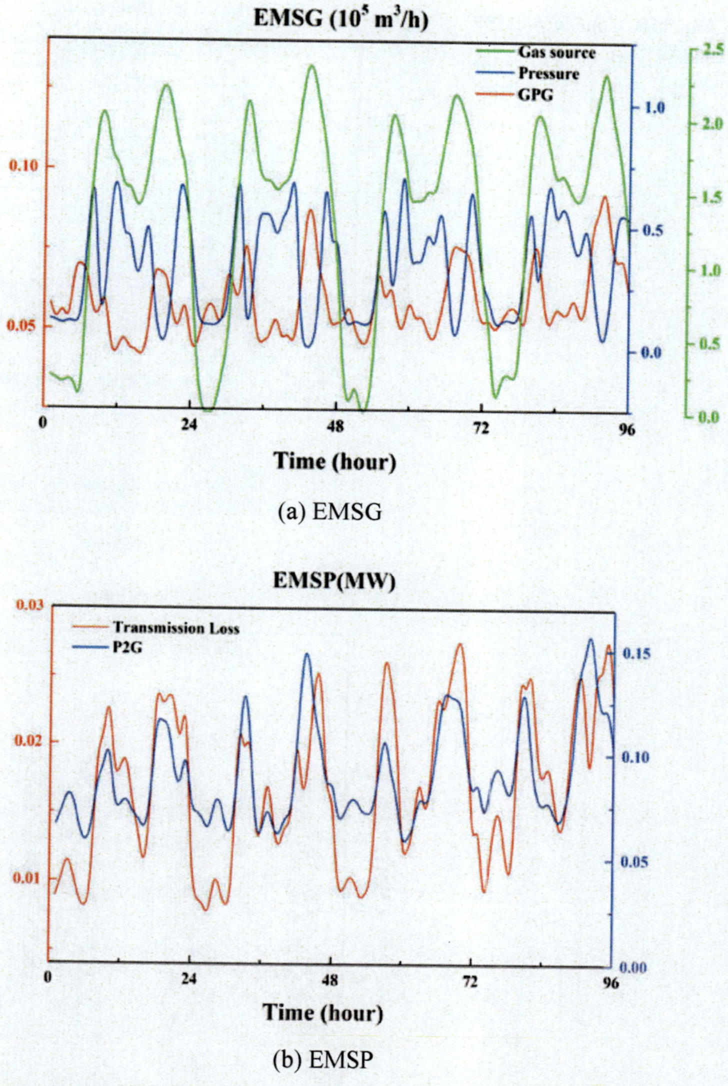

Fig. 9.9 EMSG and EMSP time series for several groupings

The EMSG/P results can be used as a crucial decision-making reference, as an adequacy index. Operators are able to determine which element is most crucial at any given moment, opposed to previous approaches, which just provide a dependability rating (as shown in Fig. 9.6). At the same time of day, EMSG and EMSP are at risk when energy demand rises, as illustrated in Fig. 9.11 and Table 9.6, respectively. Table 9.6 demonstrates that the energy source has the greatest impact on the EMSG

Fig. 9.10 Combined EMSP and EMSG

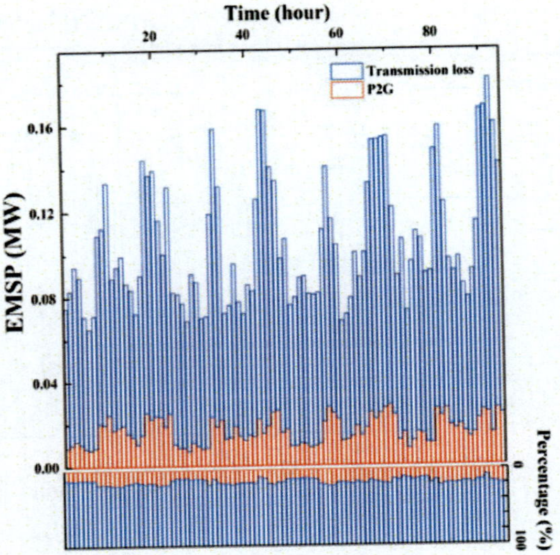

(a) EMSP

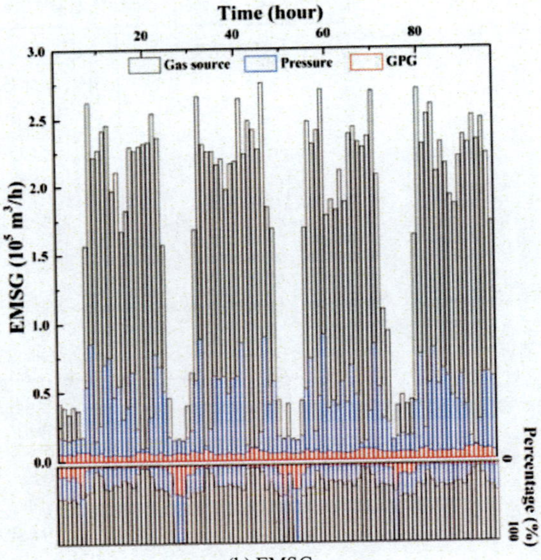

(b) EMSG

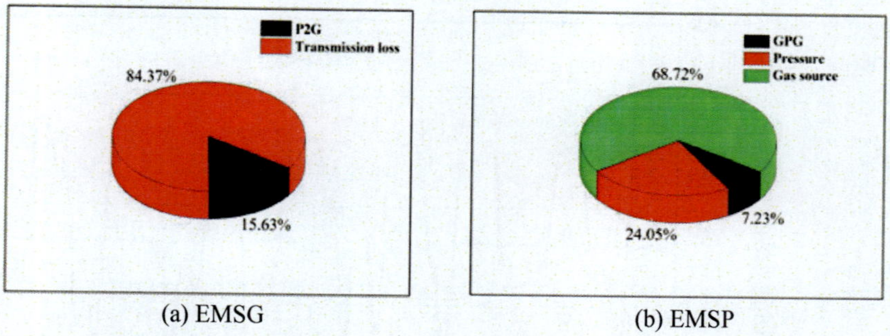

Fig. 9.11 Combined EMSG/P's percentages for several indicators

(as shown in Fig. 9.12). The results demonstrate how important GPG, P2G, and delivery pressure are to the daily energy supply.

Additionally, outcoming of forecasting, this graph might offer operators guidance. For example, as indicated in Table 9.6, the danger points in the integrated EMSG/P are different every day. This implies that the operators must make judgments in real time based on the updated index because of the variable weather and unexpected human behavior.

(B) Security assessment

In Fig. 9.12, outcomes of the relative score (RS) are shown. The value of RS reflects how distant and how close the system is to each of its functional states at any given time. The evolution of RS through time corresponds to the evolution of system functioning. RS may be calculated using the appropriate indicators, and it monitors how the system's dependability changes over time.

Table 9.6 Merged EMSG and EMSP's risk factors

	Day	Risk time (hour)	Value
EMSG (10^5 m^3/h)	1	8, 11, 12, 22	2.65, 2.45, 2.49, 2.58
	2	8, 17, 19, 22	2.70, 2.67, 2.55, 2.79
	3	8, 11, 22	2.51, 2.74, 2.73
	4	8, 10, 11, 20, 22	2.75, 2.56, 2.63, 2.58, 2.54
EMSP (MW)	1	10, 18, 20	0.088, 0.096, 0.092,
	2	9, 19, 20	0.10, 0.10, 0.11,
	3	9, 19, 20, 21, 22	0.095, 0.10, 0.10, 0.10, 0.10
	4	8, 9, 18, 19, 20, 21	0.10, 0.1, 0.11, 0.11, 0.11, 0.11

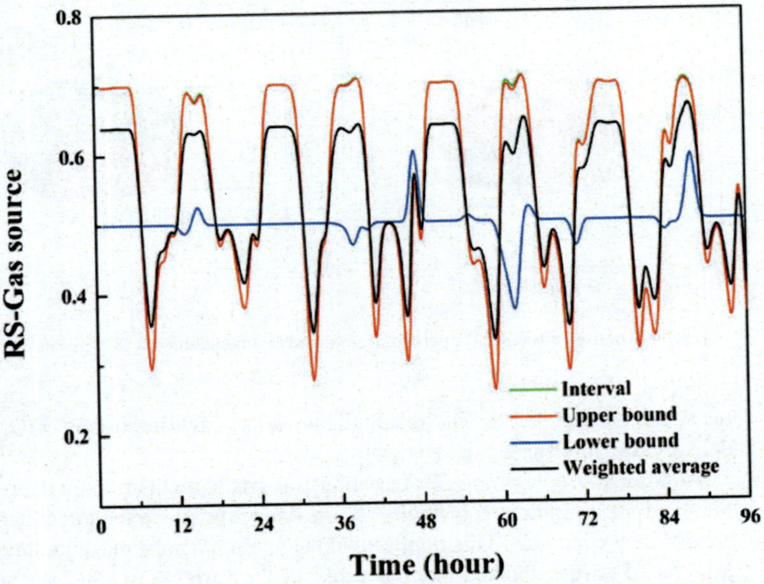

(a) RS of gas source

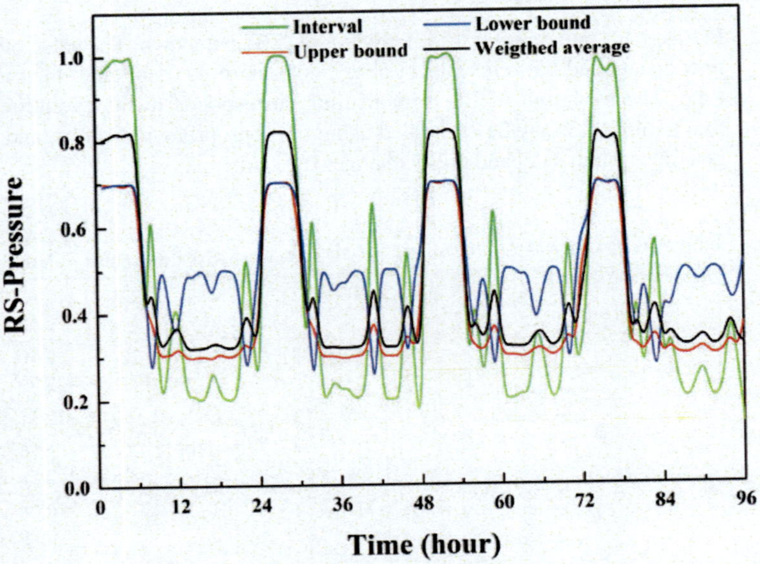

(b) RS of pressure

Fig. 9.12 RSs of different variables

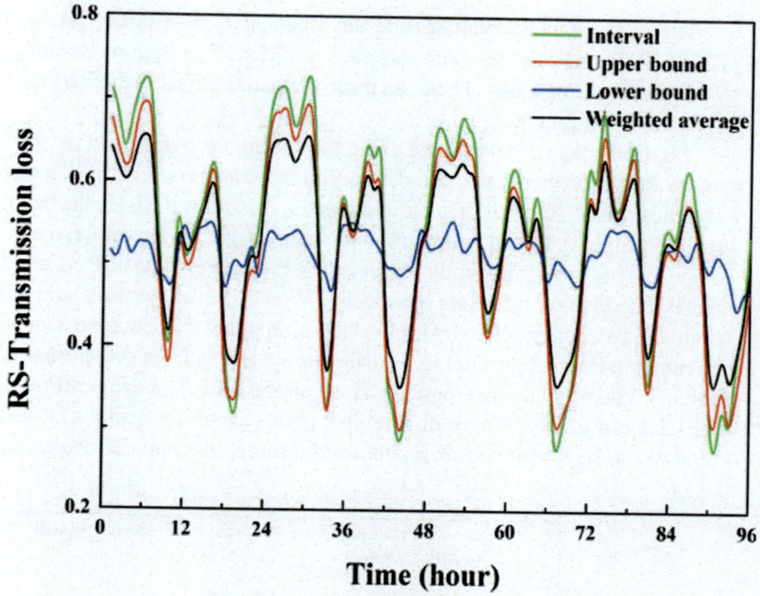

(c) RS of transmission loss

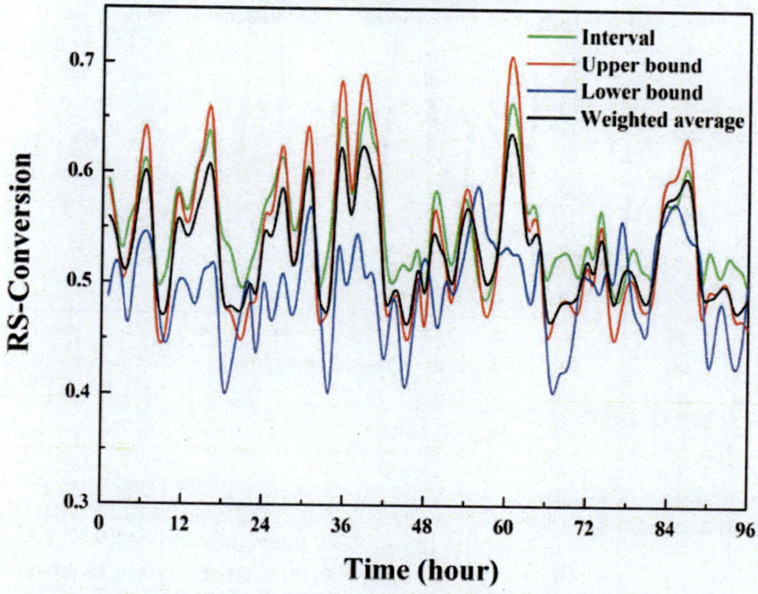

(d) RS of conversion

Fig. 9.12 (continued)

The patterns of RS demonstrate the identical circumstances when compared to EMSG/P. For nearly the entire time, when EMSG/P is high or low, the high or low value of RS corresponds. There are more plateaus and flat in the RS of the gas source and delivery pressure.

The CDI may be determined using the formula provided in Fig. 9.13, according to the RS. This shows that the delivery demand affects security dependability in a significant way. According to the conversion, the value of RS (in the danger period) is still very high (0.46–0.49), and it should be highlighted. Therefore, the dependability of security is not significantly impacted by the conversions of P2G and GPG.

As the value of indicators approaches WorstMV, the system's functioning deteriorates. When compared to EMSG/P, CDI is an index that can represent actual system circumstances by determining how far the system is from its optimal functionality. Table 9.7 displays the risk periods based on the CDI. The danger times according to the CDI and EMSG/P have distinct risk points (Tables 9.6 and 9.7). The most risky time would be when the risk points are the same, like hour 22 on the third day.

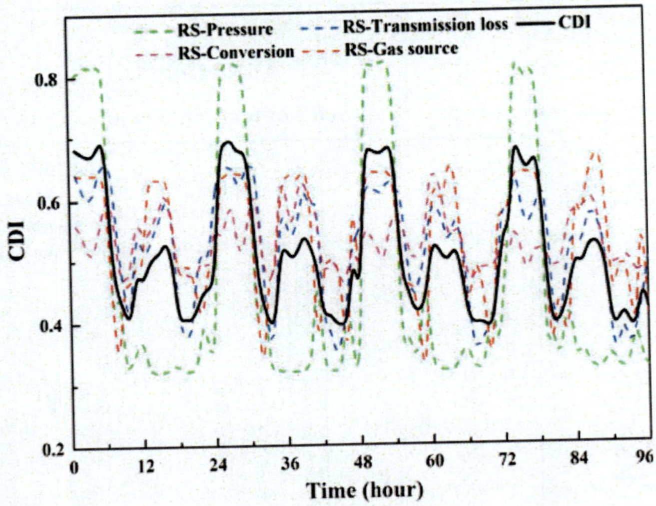

Fig. 9.13 Results of CDI

Table 9.7 Risk times according to the CDI

Day	Risk time (hour)	Value of CDI
1	9, 10, 18, 19, 20	0.42, 0.40, 0.40, 0.41, 0.40
2	9, 10, 18, 20, 21, 22	0.40, 0.40, 0.40, 0.39, 0.40
3	9, 10, 18–22	0.42, 0.41, 0.39–0.40
4	8, 9, 18, 19, 21, 22	0.39, 0.39, 0.39, 0.39, 0.39, 0.39

9.4 Conclusions

A brand-new dependability assessment approach for IES is created in this chapter. The proposed technique combines four steps: data forecasting, system functional states analysis, reliability evaluation, and IES dynamic performance simulation. Discussion and analysis follow the outcomes of the created approach on the test case.

When using the BELM technique, it is possible to quickly and accurately provide probabilistic forecasting results by estimating the PIs of demand and renewable energy output. Stack autoencoder models are utilized to create the dynamic simulation model of IES rather than conventional physical models since the simulation periods may be quite lengthy owing to the resampling from the PIs of data. These models can be good in terms of the demand for accuracy and speed of computation.

Finally, a model for system functional states analysis is proposed to defeat the restrictions of binary values in the system dependability evaluations. This model integrates the GMM-HMM methodology with OS method. Process variables are included in the model to dynamically evaluate system functioning and to provide a probability distribution with the appropriate percentile value for each condition. In comparison to conventional point values, the outcomes are more reliable and readily available.

References

1. Q. Zeng, J. Fang, J. Li, and Z. Chen, "Steady-state analysis of the integrated natural gas and electric power system with bi-directional energy conversion," *Applied Energy,* vol. 184, pp. 1483–1492, 2016, doi: https://doi.org/10.1016/j.apenergy.2016.05.060.
2. Y. Wang, J. Cheng, N. Zhang, and C. Kang, "Automatic and linearized modeling of energy hub and its flexibility analysis," *Applied Energy,* vol. 211, pp. 705–714, 2018/02/01/ 2018, doi: https://doi.org/10.1016/j.apenergy.2017.10.125.
3. M. Qadrdan, J. Wu, N. Jenkins, and J. Ekanayake, "Operating strategies for a GB integrated gas and electricity network considering the uncertainty in wind power forecasts," *IEEE Transactions on Sustainable Energy,* vol. 5, no. 1, pp. 128–138, 2013, doi: https://doi.org/10.1109/TSTE.2013.2274818.
4. X. Fu, G. Li, X. Zhang, and Z. Qiao, "Failure probability estimation of the gas supply using a data-driven model in an integrated energy system," *Applied Energy,* p. 232, 2018, doi: https://doi.org/10.1016/j.apenergy.2018.09.097.
5. E. Zio, M. Delfanti, L. Giorgi, V. Olivieri, and G. Sansavini, "Monte Carlo simulation-based probabilistic assessment of DG penetration in medium voltage distribution networks," *International Journal of Electrical Power & Energy Systems,* vol. 64, pp. 852–860, 2015, doi: https://doi.org/10.1016/j.ijepes.2014.08.004.
6. X. Fu, G. Li, and H. Wang, "Use of a second-order reliability method to estimate the failure probability of an integrated energy system," *Energy,* vol. 161, pp. 425–434, 2018, doi: https://doi.org/10.1016/j.energy.2018.07.153.
7. H. Su *et al.*, "A systematic method for the analysis of energy supply reliability in complex Integrated Energy Systems considering uncertainties of renewable energies, demands and operations," *Journal of Cleaner Production,* vol. 267, p. 122117, 2020/09/10/ 2020, doi: https://doi.org/10.1016/j.jclepro.2020.122117.

8. D.-A. Ciupăgeanu, G. Lăzăroiu, and L. Barelli, "Wind energy integration: Variability analysis and power system impact assessment," *Energy*, vol. 185, pp. 1183–1196, 2019, doi: https://doi.org/10.1016/j.energy.2019.07.136.
9. S. Clegg and P. Mancarella, "Integrated Modeling and Assessment of the Operational Impact of Power-to-Gas (P2G) on Electrical and Gas Transmission Networks," *IEEE Transactions on Sustainable Energy*, vol. 6, no. 4, pp. 1234–1244, 2015, doi: https://doi.org/10.1109/TSTE.2015.2424885.
10. A. Alabdulwahab, A. Abusorrah, X. Zhang, and M. Shahidehpour, "Coordination of interdependent natural gas and electricity infrastructures for firming the variability of wind energy in stochastic day-ahead scheduling," *IEEE Transactions on Sustainable Energy*, vol. 6, no. 2, pp. 606–615, 2015, doi: https://doi.org/10.1109/tste.2015.2399855.
11. P. Pinson and G. Kariniotakis, "Conditional Prediction Intervals of Wind Power Generation," IEEE Transactions on Power Systems, vol. 25, no. 4, pp. 1845–1856, 2010, doi: https://doi.org/10.1109/tpwrs.2010.2045774.
12. NERC. "Integrated reliability index (IRI) concepts—comments and responses." https://www.nerc.com/comm/PC/PerformanceAnalysisSubcommitteePASDL/CDI_Whitepaper.pdf accessed.

Chapter 10
Fault Detection and Diagnose Method for Pressurization Devices

Lin Fan and Shiliang Peng

Abstract Planners may focus on the health of the equipment using the condition-based maintenance technique and make recommendations based on the data collected. However, the effectiveness of the fault detection and diagnosis technique must be established first. The safety and dependability of oil and gas pipeline systems may be considerably improved by fault detection of pressurization devices. A deep learning-based fault detection and diagnosis approach is introduced in this chapter. The diagnostic accuracy for flaws of this technology, which combines time–frequency domain analysis and visual pattern recognition algorithms, can be above 95%.

10.1 Safety of Oil and Gas Pipeline System

With the rapid development of the world's low carbon economy, the demand for natural gas as an efficient fossil energy source is showing a high growth trend. As the bond between natural gas resources and downstream markets, the natural gas pipeline system is the main way to secure the demand of natural gas users. Therefore, the ability of the natural gas pipeline network (NGPN) system to operate safely and reliably determines the safety and security of basic energy facilities [1, 2]. In this regard, studies have looked at the effects and likelihoods of NGPN unit failures on the network's dependability and safety [3–6]. The following are two types of techniques to assess or enhance the reliability and security of NGPN system provisioning: (1) analysis of system-level probabilistic security (PSA) and (2) unit-level prognostic health management (PHM).

L. Fan (✉)
PetroChina Planning & Engineering Institute, 100083 Beijing, China
e-mail: fanlin927@126.com

S. Peng
National Engineering Laboratory for Pipeline Safety; Beijing Key Laboratory of Urban Oil and Gas Distribution Technology, China University of Petroleum, Beijing 102249, China
e-mail: Andypeng215@outlook.com

H. Su et al. (eds.), *Advanced Intelligent Pipeline Management Technology*,
https://doi.org/10.1007/978-981-19-9899-7_10

PSA can evaluate the whole range of potential scenario outcomes and offer recommendations for risk-management tactics [7]. It is a useful tool for managers to comprehend prospective accidents, their likelihood, and their effects. A capacity network stochastic model is created by Su et al. [6] to quantitatively evaluate supply dependability for NGPN systems. Different customers' gas shortage risk is examined from both a global and a personal standpoint.

An effective supply reliability evaluation approach that enables to investigate the relationships between failure events and shortage accidents is required for safeguarding the system's energy supply. A well-liked probabilistic technique for predicting system dependability and assisting in decision-making is the Bayesian network (BN) [8, 9]. The application of BNs in assessing the system reliability was summarized by Adedipe [8]. Song et al. [10] proposed a dependence limit analysis approach to quantify the reliability of a complex system using the probabilistic modeling.

Although previous work on optimization of system reliability issues in pipeline system has progressed, it is still an unresolved topic. First, the computational workload of current methods for modeling gas supply reliability in natural gas pipeline networks grows exponentially with the magnitude of the problem. Second, the same maintenance intervals are currently used for similar installations in the oil and gas industry system, although the rate of installation degradation varies depending on the operating environment. This could result in some units receiving excessive maintenance, while others receive insufficient maintenance. Third, the actual maintenance of each unit in the NGPN system is only related to its current operational status without considering its importance to the system function.

10.2 Description of the Mathematical Model

This paper presents a preventive maintenance strategy optimization method for gas transmission pipeline systems based on gas supply reliability. The method consists of three parts: a mapping relationship model between maintenance strategy and unit reliability, a fast evaluation method of system gas supply reliability based on maintenance strategy, and a dynamic optimization method of maintenance strategy based on deep reinforcement learning. This part gives a description of the optimization of supply reliability for natural gas pipeline system while also taking into account the components (units). Figure 10.1 shows a general summary of the established technique.

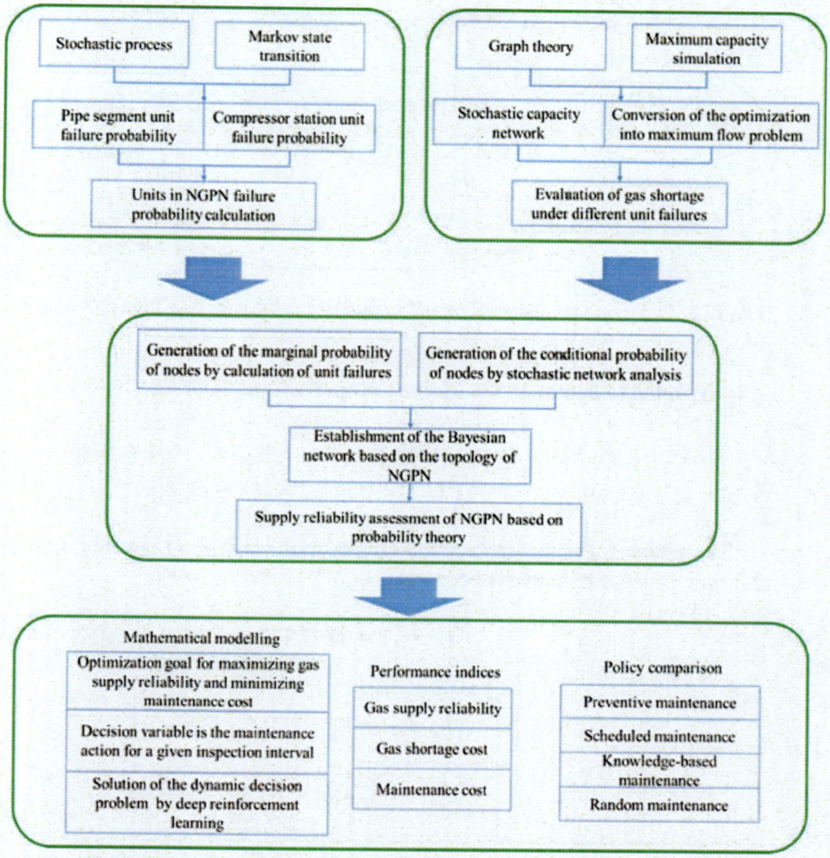

Fig. 10.1 Description of the method

10.2.1 Modeling the Pipeline System

The pipeline system is made up of several units of various sorts that are linked together by intricate logical linkages. Pipelines, compressors, gas providers, and CUs are a few of these units. The units modeling will be described in the paragraphs that follow.

Different operational states of the units are defined as a result of complex operating conditions and outside environmental disturbances. Figure 10.2 shows the state change mechanism for these units. The pipelines are classified as being in normal, deteriorated, and interrupted (failure) states.

The three states normal, deteriorated, and failure are also specified for the compressor stations (Fig. 10.2). The term "degenerated condition" refers to a compressor failure that results in a 20% drop in the capacity of the associated pipeline. We utilize the Markov stochastic process to explain the change of the units' state as

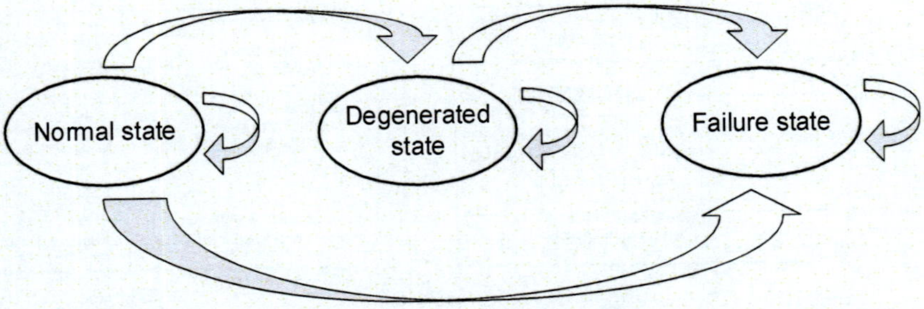

Fig. 10.2 Description of the pipeline and compressor units' state transition procedure

it is expected to adhere to the Markov property.

$$\begin{aligned} P\{X(s+t) = j|X(s) = i,\ X(u) = x(u),\ 0 < u < s\} \\ = P\{X(s+t) = j|X(s) = i\} \end{aligned} \tag{10.1}$$

The above formula can be abbreviated as $P_{i,j}(s, s+t)$, and the homogeneous Markov process hypothesis becomes:

$$P_{i,j}(s, s+t) = P_{i,j}(t) \tag{10.2}$$

$$\boldsymbol{A} = \begin{bmatrix} -\sum\limits_{j=1}^{n} \alpha_{0j} & \alpha_{01} & \cdots & \alpha_{0n} \\ \alpha_{10} & -\sum\limits_{\substack{j=1 \\ j\neq 1}}^{n} \alpha_{1j} & \cdots & \alpha_{1n} \\ \vdots & \vdots & \ddots & \vdots \\ \alpha_{n0} & \alpha_{n1} & \cdots & -\sum\limits_{j=1}^{n-1} \alpha_{nj} \end{bmatrix} \tag{10.3}$$

$$\begin{cases} \dfrac{\mathrm{d}\boldsymbol{P}}{\mathrm{d}t} = [P_1(t), P_2(t), \ldots, P_n(t)] \cdot \boldsymbol{A} \\ \boldsymbol{P}(t_0) = \boldsymbol{C} \end{cases} \tag{10.4}$$

Figure 10.2 represents the state transition diagram for a pipe of length L. The transition rate matrix is shown in Eq. (10.3). The following matrix illustrates the unit state changeover procedure:

$$\boldsymbol{A} = \begin{bmatrix} -(\lambda_{P1} + \lambda_{P3}) \cdot L & \lambda_{P1} \cdot L & \lambda_{P3} \cdot L \\ 0 & -\lambda_{P2} \cdot L & \lambda_{P2} \cdot L \\ \mu \cdot L & 0 & -\mu \cdot L \end{bmatrix} \tag{10.5}$$

Typical sources of gas for NGPN are upstream pipelines, LNG units, or underground storage facilities. Numerous factors, such as those related to the economy, the environment, and politics, have an impact on the stability of gas supplies. We examine the two states of normal and failure to characterize the gas source in order to keep things simple without sacrificing generality.

The customer status is determined by the amount of gas demanded by the customer in conjunction with the actual amount of gas supplied. According to the equilibrium between supply and demand, three states are used to describe the customer state: normal, shortage, and severe shortage. Normal means that the supply is equal to the customer's demand; shortage means that the actual supply can partially meet the customer's demand; and severe shortage means that the actual supply is less than 75% of the demand. In addition, the gas source of a natural gas pipeline is usually an oil and gas field, the end of a long-distance pipeline or a gas storage reservoir, where the gas source is always in a normal state without considering the failure of the gas source due to an unexpected accident. In practice, the determination of the division threshold or the number of states for the user state needs to be further analyzed in relation to the specific problem scenario.

10.2.2 Supply Capacity Calculation

10.2.2.1 State Transition Modeling

We developed a graph theory-based capacity network representation that characterizes the operational state of the system and highlights the complex topology of the NGPN system. The weights of the edges in the capacity network represent the pipes' maximum supply capacity and depend on their condition.

The values of edge weights are recorded by the matrix Q, while the distances between linked nodes are recorded using the lengths matrix L. When a network component malfunctions, the capacity matrix is altered, and the pipeline's capacity between nodes i and j is noted as $Q\ (i, j)$.

10.2.2.2 Stochastic Capacity Network

To ensure gas supply reliable and improve system efficient, the operator is recommended to deliver the natural gas based on flow direction while transitioning between unit states. We provide a shortest-path strategy for the redirection that integrates maximum flow for a random network model with a priority pattern in order to reproduce the optimal operating environment for the NGPN system.

In order to maximize the amount of material that may be delivered between two sources and sinks with a restricted capacity, the optimal utilization of line capacity is outlined in terms of the maximum flow problem. In NGPN systems, the maximum flow algorithm can be used to determine the supply capacity [11–13].

When the state of natural gas pipeline equipment changes, the distribution of flow in the pipeline needs to be readjusted to maximize the gas demand of downstream users. Therefore, the maximum flow algorithm is introduced to calculate the maximum gas supply capacity of a natural gas pipeline system under different operating conditions [13]. The maximum flow algorithm transforms the gas pipeline flow allocation problem into a calculation of the maximum delivery capacity at two points in the transmission network, where the gas source and the customer are considered as the source and sink points, respectively. For natural gas pipeline systems where there are multiple gas sources or multiple users, the multiple source and multiple sink problem needs to be transformed into a single source and single sink problem. Therefore, we build on the original topology by introducing a virtual source point U and a virtual consumer point V. The virtual source point is connected to all sources and supplies gas to all sources. The virtual gas consumption point is connected to all users, and the weight of the connected edge is the gas demand of the user.

10.2.3 Probabilistic Model for Pipeline System

The analysis of gas supply reliability of pipeline systems using Bayesian networks involves four key steps: (1) combining expert experience with pipeline operational data to determine the Bayesian network variable nodes as well as the network structure. (2) Combining the unit life distribution model to calculate the unit instantaneous failure probability as input to the Bayesian network node edge distribution. (3) Combine the maximum flow algorithm with Monte Carlo to simulate the user gas shortage state under different unit failures, and use parameter learning to determine the Bayesian network node conditional probability distribution. (4) Using Bayesian inference to calculate the probability distribution of user nodes, and establish the evaluation of gas supply reliability of the pipeline system.

System reliability evaluation has involved a number of conventional reliability models. However, the potential for their implementation for NGPN system stability is limited by the numerous NGPN units, each of which has several states and is interconnected via complicated network topologies. The Bayesian theorem summarizes the probability relationship between unit failures and gas shortages as follows:

$$P(y|x) = \frac{P(x|y) \cdot P(y)}{P(y)} \tag{10.6}$$

where variable x indicates a failure event of component and variable y indicates the event of gas shortage.

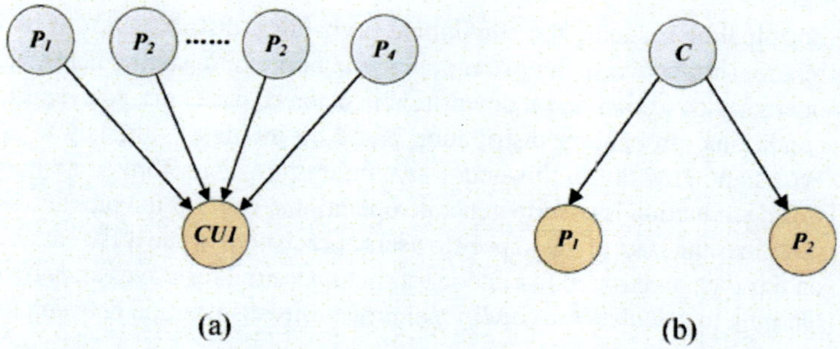

Fig. 10.3 Bayesian network for a simple pipeline system

10.2.3.1 Variables Definition

The domain values are shown as nodes, while the network graph's dependencies are encoded. The BN's structure explains the causal connections between nodes and incorporates the system's inherent dynamics.

Two Bayesian models for a simple pipeline system are depicted in Fig. 10.3. The input of the root leaf is the marginal probability of pipeline failure. The leaf node indicates the supply capacity of components (pipeline and compressor). The failure of the pipeline is indicated by node *P*, and the failure of the pressure station is indicated by node *C*. The gas shortage situations of the CUs are represented by the CU node. Figure 10.3a illustrates the scenario in which the linked pipes degenerate as a result of the compressor station failing. Gas shortages for the downstream CUs are caused by a reduction in supply capacity.

10.2.3.2 Marginal Probability Distribution

Random failures of key units (compressors and pipe sections) within a natural gas pipeline system can lead to gas shortages for downstream customers, and the use of Bayesian networks can describe the causal relationships between variables. Bayesian networks are typically directed acyclic graphs, consisting mainly of nodes and directed edges. The nodes represent the random variables, the directed edges represent the interrelationships between the nodes, and the strength of the interaction between the two variables is expressed as a conditional probability. The structure of Bayesian networks intuitively reflects the causal logic between variables and can clearly describe the propagation of unitary uncertainty in a pipeline system. This section estimates the parameters of the BN model by Bayesian estimation:

$$p(y_i|x) = \frac{p(y_i)p(x|y_i)}{\sum_{n}^{i=1} p(y_i)p(x|y_i)} \tag{10.7}$$

For pipeline systems, the conditional probability distribution reflects the mapping relationship between an upstream pipeline or pressure station being in an abnormal operating condition and a downstream customer gas shortage event. Obtaining the conditional probability distribution based on the data is the key to constructing a Bayesian network. To this end, a maximum flow algorithm combined with Monte Carlo simulation is used to generate operational data for the pipeline system, which describes the state of gas supply to users at each node when different units fail. Based on this data, maximum likelihood estimation is used for Bayesian network parameter learning to calculate the conditional probability distribution between unit nodes and user nodes.

10.2.3.3 Index of System Supply Reliability

The probability of gas shortage can be calculated based on the probability of failure of the unit. When new supporting data become available, the powerful method of Bayesian inference may be used to evaluate the likelihood of an occurrence. The likelihood of a gas shortage at the CU at time t may be estimated using an approximation of the belief propagation technique. The indexes of system supply reliability are described as follows:

$$R_{\text{customer},i} = 1 - \frac{\sum_{t=1}^{T} P_{i,t}}{T} \tag{10.8}$$

$$R_{\text{system},t} = 1 - \frac{\sum_{i=1}^{N} P_{i,t}}{N} \tag{10.9}$$

where $R_{\text{customer},i}$ indicates the individual customers' reliability of gas supply. $R_{\text{system},t}$ indicates the system reliability of gas supply. $P_{i,t}$ is the probability of gas shortage event happening.

Indexes of adequacy are given as:

$$S_{\text{customer},i} = \frac{\sum_{t=1}^{T} P_{i,t} \cdot (D_{i,t} - S_{i,t})}{T} \tag{10.10}$$

$$S_{\text{system},t} = \frac{\sum_{i=1}^{N} P_{i,t} \cdot (D_{i,t} - S_{i,t})}{N} \tag{10.11}$$

where $S_{\text{customer},i}$ denotes the amount of natural gas unsupplied by gas source from the customer perspective. $S_{\text{system},t}$ denotes the amount of natural gas unsupplied by gas source from the system perspective. $D_{i,t}$ and $S_{i,t}$ denote the natural gas supplied and demanded to CU i at discrete time t.

10.3 Maintenance Optimization Based on RL

In this work, the Markov decision process (MDP) is adopted here to represent the dynamic maintenance optimization issue, and the sequential decision problem is solved by reinforcement learning (RL).

The architecture for preventative maintenance for the NGPN system is shown in Fig. 10.4, which combines the capability of BNs for causal inference with intelligent online learning process.

10.3.1 Modeling the Maintenance Optimization Problem

The NGPN system serves as the agents' environment in this part. The maintenance strategy is developed by the strong learning ability of agents. Addressing this finite MDP means searching a policy to achieve large returns in the long term; let $V^{\pi}(s)$ denote the state-value function:

$$V^{\pi}(s) = E_{\pi}\left[\sum_{k=0}^{\infty} \gamma^{k} r_{t+k} | S_t = s\right] \tag{10.12}$$

The above equation can be decomposed into the Bellman equation [13]:

$$V^{\pi}(s) = \sum_{a} \pi(a|s) \sum_{s',r} p(s', r|s, a)\left[r + \gamma V^{\pi}(s')\right] \tag{10.13}$$

The major goal of this part is to optimize the maintenance interval in order to lower maintenance costs. Thus, the maintenance expense is represented by an aggregate sum. Immediate rewards are defined by losses consisting of gas shortage costs and maintenance costs, as defined below:

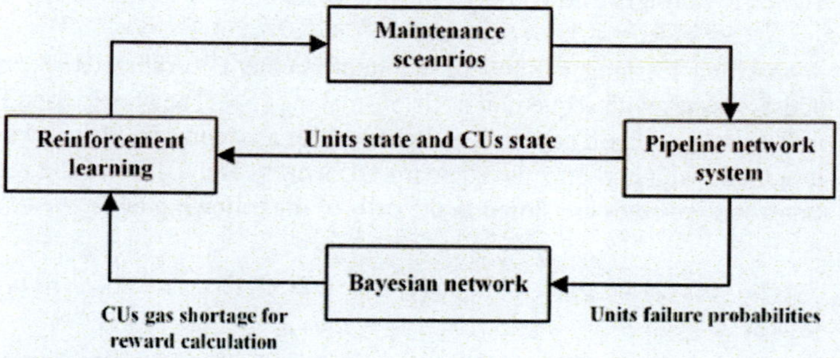

Fig. 10.4 Interaction between reinforcement learning and pipeline system

$$R(s_t, a_t, s_{t+1}) = -\sum_{t=1}^{T}\left(\alpha \cdot S_{\text{system},t} + \beta \cdot \sum_{n=1}^{N} a_{n,t} c_{R,k}\right) \tag{10.14}$$

where $c_{R,k}$ denotes the cost of each maintenance action. It is inevitable that equipment in pipeline systems will deteriorate in performance and break down during service. The maintenance management for equipment depends on the type of failure and failure mode of the equipment. In the case of pipe sections, for example, the causes of failure include material failure, third-party damage, corrosion perforation and personnel mishandling, which ultimately leads to incidents such as leaks, perforations and fractures in the pipeline. In order to reduce the risk of corrosion failure of pipe sections, maintenance can be adopted, including regular internal testing and defect repairing. Environmental monitoring around the pipeline should be enhanced for third-party damage and human malfunction. Regulations should be publicized to improve the skills of operators. As we can see, equipment failures may be caused by a variety of factors, and specific maintenances need to be further developed in conjunction with specific fault information for a particular fault type.

In practice, equipment is in a dynamic, complex interaction environment with multiple risks overlapping during operation, making it difficult to use a particular or several failure modes to describe the failures that can occur during the equipment's life cycle. For this reason, the maintenance proposed in this paper is not a specific maintenance action for a particular type of failure, but represents a series of restorative actions to restore equipment from an abnormal state to its original state. The maintenance costs incurred are comprehensive, representing the cost of equipment, manpower, time, and technology resulting from the maintenance activity. Based on this, this paper focuses on improving pipeline system reliability by optimizing the maintenance time point of equipment.

10.3.2 DRL-Based Maintenance Management Framework

10.3.2.1 Integrating RL with Maintenance

A machine learning method called reinforcement learning (RL) can effectively address issues with sequential decision-making [14]. The widely used RL algorithm is Q-learning (QL). The action and state spaces are represented by rows and columns in a table, which is how the standard QL works. The values of the related pairs of state–action values are stored in the cells of the following table:

$$Q_{k+1}(s_t, a_t) = Q_k(s_t, a_t) + \alpha\left(r_t + \gamma \max_a Q_k(s_{t+1}, a) - Q_k(s_t, a_t)\right) \tag{10.16}$$

where α represents the learning rate, which determines the iteration process of the algorithm.

10.3.2.2 Maintenance Optimization Based on DRL

To overcome the above problems, we propose a DRL-based maintenance planning strategy for NGPN systems under random failures. Two neural networks are constructed, each with the same structure but different parameters, to lessen the instability brought on by the correlation of data collected at various periods. Updated policies are intended to reduce function loss, which is expressed as follows:

$$l(\theta) = \mathbb{E}\left[\left(r + \gamma \max_{a_{t+1}} Q\left(s_{t+1}, a_{t+1}, \theta^{-}\right) - Q(s_t, a_t, \theta)\right)^2\right] \tag{10.17}$$

θ and θ^{-} are parameters for two neural networks sharing the same parameters. The parameters are updated for fixed episodes.

A collection of training data may be used to compute the function loss. The training dataset is chosen as a sample from a buffer known as experience replay. The main targets of empirical replay techniques are data correlation problems and non-smooth distribution of empirical data [15]. Figure 10.5 depicts the policy's neural network structure.

10.4 Validation for the Proposed Method

10.4.1 Parameters for Pipeline System

This section conducts numerical tests to evaluate the effectiveness of the proposed technique leveraging the data in references [16]. The topology of the tested natural gas pipeline system is shown in Fig. 10.6. The NGPN is 692.8 km long in its whole. In the NGPN under test, there are 2 gas compressor stations, 2 gas suppliers, and 8 end consumers. The maximum operation pressure of the pipeline is 10 MPa; it has an outside diameter of 1067 mm and a wall thickness of 12.5 mm. The pipe segment's average failure rate is calculated at 3.5×10^{-4} per kilometer-year.

The NGPN systems' most common maintenance technique in use today is planned maintenance. This article merely introduces the random maintenance method as a hypothetical situation that represents the worst-case scenario of absence experience and accompanying management strategies. Simply said, various maintenance techniques are measured against the random maintenance plan.

- Preventive maintenance strategy: carry out repair tasks on all units while taking into consideration the expense of a gas shortage and the cost of maintenance in the pipeline system. Use information about CU's natural gas shortage potential and system maintenance costs when making decisions.
- Scheduled maintenance strategy: use regular maintenance intervals to maintain the piping and compressor until the end of the maintenance period, then make

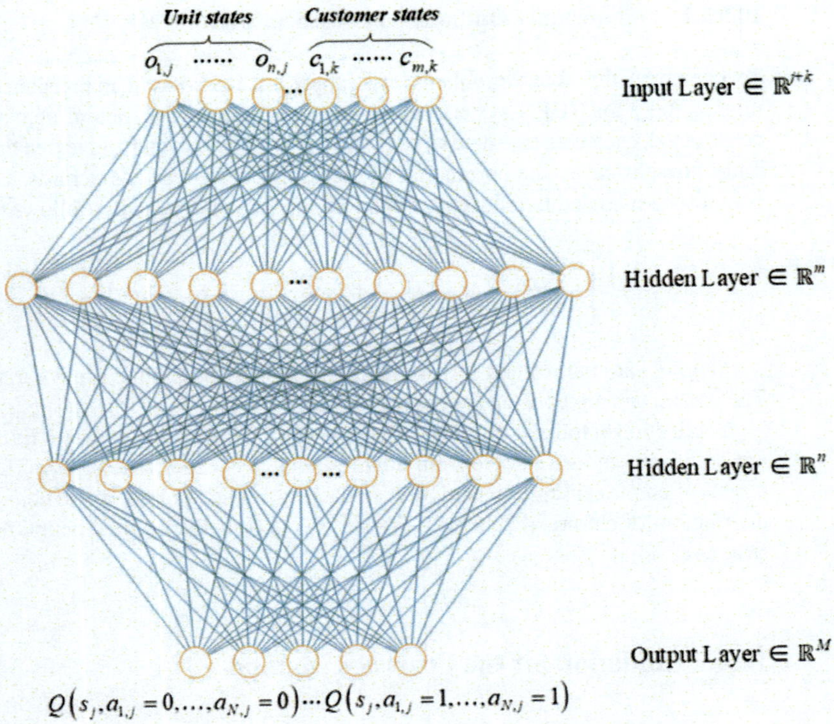

Fig. 10.5 Neural network structure

the necessary repairs. The predicted system's health status is used to estimate the inspection interval.

- Knowledge-based maintenance strategy: carry out the necessary maintenance on the important units. In the network shown along Fig. 10.6, the important units are those that are in the main path of the network or are close to the gas source, such as pipe segments 1–3, 3–5, and 2–12.
- Random maintenance strategy: randomly carry out maintenance action on gas compressors and pipelines. If a device malfunctions while it is in use, it will be repaired and returned to its original condition.

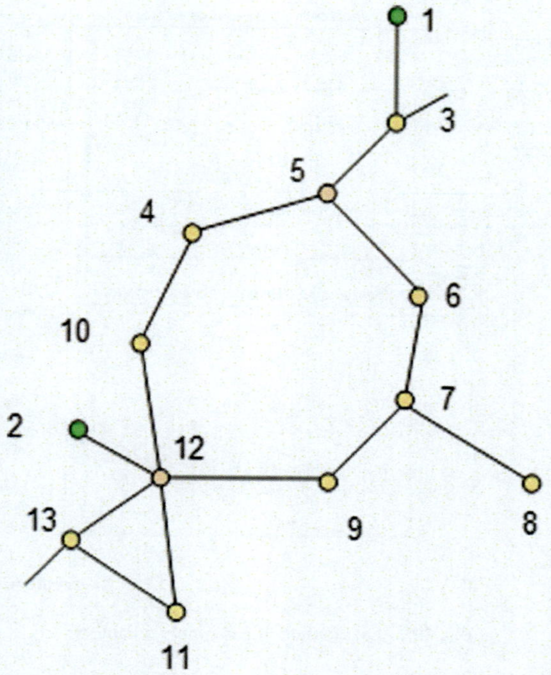

Fig. 10.6 Topology structure of the pipeline system

10.4.2 Results Analysis

10.4.2.1 Comparison of Gas Supply Reliability

Figure 10.7 displays the outcomes of the 48-month study for CUs under different maintenance strategies. The empirical probability distribution functions for the reliability of the CU gas supply under various technical conditions are shown by bar histograms, and the empirical cumulative distribution functions are indicated by red lines. The comparison of the findings shows that both random and preventative maintenance are helpful in ensuring operation safety and system efficiency. Besides, the efficacy of the maintenance strategy depends on both the estimated cost of a gas shortage and the cost of upkeep. The former is calculated by multiplying the likelihood of a gas shortage by its impact. So it should come as no surprise that random maintenance might also result in a highly reliable system. Good system costs can nevertheless be a result of severe effects but high dependability. In Sect. 10.4.2.2, several maintenance techniques' costs related to gas shortages are compared.

The analysis of system-level gas supply reliability for various maintenance methods is shown in Fig. 10.8, where the average value of CU gas supply reliability

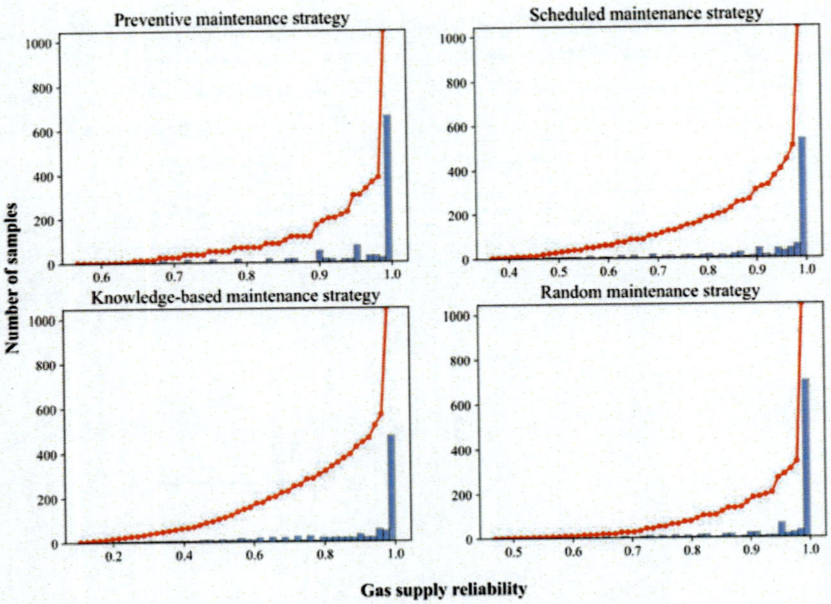

Fig. 10.7 Distribution of gas supply reliability

at each time point is the system gas supply reliability. The density function represented in Fig. 10.9 further indicates that the methods such as preventive maintenance methods and the stochastic maintenance methods achieve higher supply reliability.

10.4.2.2 Cumulative Gas Shortage

Figure 10.10 shows the outcomes of the suggested maintenance method. System performance is defined in terms of the likelihood of shortage occurrences occurring by gas supply reliability. Equations (10.8)–(10.11) are used to calculate the amount of natural gas shortage.

10.5 Conclusion

This chapter develops a unique system management framework for natural gas pipeline network (NGPN) systems that combines reliability optimization and preventive maintenance for pipeline system. Instead of considering the operating conditions and health of the units, maintenance plans are created with reference to data on the likelihood of a gas shortage and the expense of system upkeep. A Bayesian

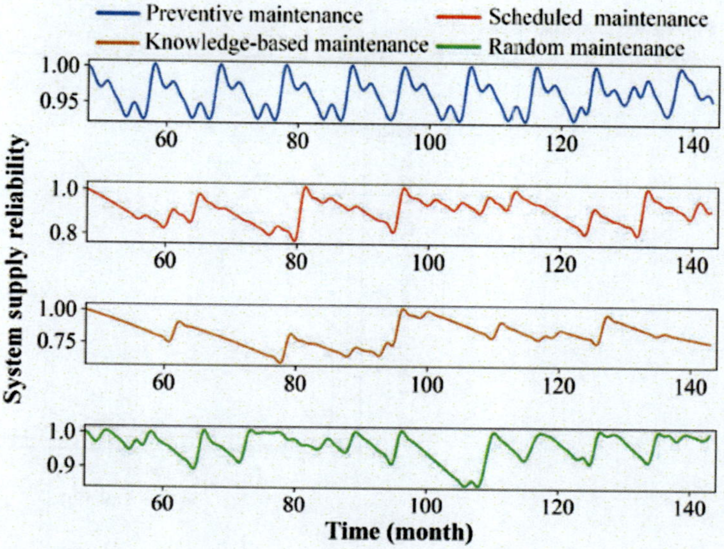

Fig. 10.8 Comparison of system supply reliability

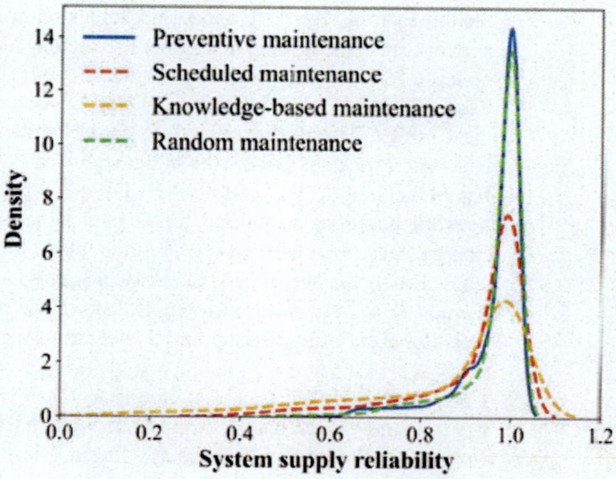

Fig. 10.9 Probability density functions for the system's gas supply reliability

network (BN)-based gas supply reliability assessment model represents the relationship between unit failure and CU gas shortage. Besides, a stochastic model network modeling is developed to generate data for maintenance management. The DRL-based maintenance strategy can guarantee system reliability and reduce maintenance cost. The proposed method can provide insight into intelligent management for energy system.

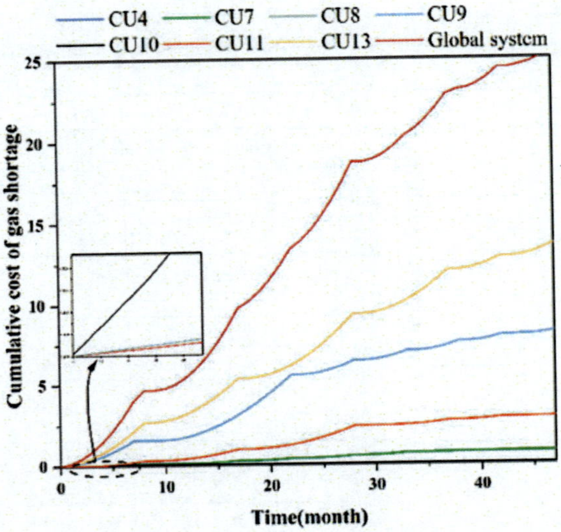

Fig. 10.10 Cumulative gas shortage

References

1. Shaikh F, Ji Q, Fan Y. Evaluating China's natural gas supply security based on ecological network analysis. Journal of Cleaner Production 2016; 139:1196–206. https://doi.org/10.1016/j.jclepro.2016.09.002.
2. Zhang T, Bai H, Sun S. Intelligent Natural Gas and Hydrogen Pipeline Dispatching Using the Coupled Thermodynamics-Informed Neural Network and Compressor Boolean Neural Network. Processes 2022; 10:428. https://doi.org/10.3390/pr10020428.
3. Fan MW, Gong J, Wu Y, Kong WH. The gas supply reliability analysis of natural gas pipeline network based on simplified topological structure. Journal of Renewable and Sustainable Energy 2017; 9. https://doi.org/10.1063/1.4997490.
4. Yu W, Huang W, Wen Y, Li Y, Liu H, Wen K, et al. An integrated gas supply reliability evaluation method of the large-scale and complex natural gas pipeline network based on demand-side analysis. Reliability Engineering and System Safety 2021; 212:107651. https://doi.org/10.1016/j.ress.2021.107651.
5. Chen Q, Zuo L, Wu C, Bu Y, Huang Y, Chen F, et al. Supply adequacy assessment of the gas pipeline system based on the Latin hypercube sampling method under random demand. Journal of Natural Gas Science and Engineering 2019; 71:102965. https://doi.org/10.1016/j.jngse.2019.102965.
6. Su H, Zhang J, Zio E, Yang N, Li X, Zhang Z. An integrated systemic method for supply reliability assessment of natural gas pipeline networks. Applied Energy 2018; 209:489–501. https://doi.org/10.1016/j.apenergy.2017.10.108.
7. de Vasconcelos V, Soares WA, da Costa ACL, Raso AL. Deterministic and Probabilistic Safety Analyses. 2019. https://doi.org/10.1016/b978-0-12-815906-4.00002-6.
8. Adedipe T, Shafiee M, Zio E. Bayesian Network Modelling for the Wind Energy Industry: An Overview. Reliability Engineering and System Safety 2020; 202:107053. https://doi.org/10.1016/j.ress.2020.107053.
9. Guo Y, Zhong M, Gao C, Wang H, Liang X, Yi H. A discrete-time Bayesian network approach for reliability analysis of dynamic systems with common cause failures. Reliability Engineering and System Safety 2021; 216:108028. https://doi.org/10.1016/j.ress.2021.108028.

10. Song Y, Mi J, Cheng Y, Bai L, Chen K. A dependency bounds analysis method for reliability assessment of complex system with hybrid uncertainty. Reliability Engineering and System Safety 2020; 204:107119. https://doi.org/10.1016/j.ress.2020.107119.
11. Ye Y, Grossmann IE, Pinto JM, Ramaswamy S. Modeling for reliability optimization of system design and maintenance based on Markov chain theory. Computers and Chemical Engineering 2019; 124:381–404. https://doi.org/10.1016/j.compchemeng.2019.02.016.
12. Li H, Zhang T, Zhang Y, Wang K, Li J. A maximum flow algorithm based on storage time aggregated graph for delay-tolerant networks. Ad Hoc Networks 2017; 59:63–70. https://doi.org/10.1016/j.adhoc.2017.01.006.
13. Şuvak Z, Altınel K, Aras N. Exact solution algorithms for the maximum flow problem with additional conflict constraints. European Journal of Operational Research 2020; 287:410–37. https://doi.org/10.1016/j.ejor.2020.04.001.
14. Mason K, Grijalva S. A review of reinforcement learning for autonomous building energy management. Computers and Electrical Engineering 2019; 78:300–12. https://doi.org/10.1016/j.compeleceng.2019.07.019.
15. Wawrzyński P, Tanwani AK. Autonomous reinforcement learning with experience replay. Neural Networks 2013; 41:156–67. https://doi.org/10.1016/j.neunet.2012.11.007.
16. Praks P, Kopustinskas V, Masera M. Probabilistic modelling of security of supply in gas networks and evaluation of new infrastructure. Reliability Engineering and System Safety 2015; 144:254–64. https://doi.org/10.1016/j.ress.2015.08.005.

Chapter 11
Intelligent Leakage Detection for Pipelines

Jian Du and Jianqin Zheng

Abstract In recent years, due to the internal and external factors, pipeline leakage accidents happen frequently which lead to hidden danger to the safe operation of the pipeline. The pipeline leakage accidents not only cause serious economic loss, but also harm the safe operation of pipeline and personal safety. Consequently, it is extremely significant to detect and locate the leakage of the pipeline in time. Existing methods of detecting and locating pipeline leakage can be divided into two types, external and internal monitoring. This chapter summarizes the common methods for pipeline leak detection and location, including acoustic methods, negative pressure waves, intelligent algorithm-based methods, and data-driven methods. The applications of different methods are also given to compare their strengths and weaknesses.

11.1 Pipeline Leakage Detection

Transporting hazardous chemicals is an extremely significant project, and pipeline transportation is the most common mode of transportation at present. Compared with other ways of transportation, pipelines are provided with many advantages. For example, it has higher transportation efficiency, and cost lower energy consumption, multiple products can be transported continuously [1]. But in recent years, due to the internal and external factors, pipeline leakage accidents happen frequently which lead to hidden danger to the safe operation of the pipeline. The pipeline leakage accidents not only cause serious economic loss [2], but also harm the safe operation of pipeline and personal safety.

J. Du
National Engineering Laboratory for Pipeline Safety/MOE Key Laboratory of Petroleum Engineering; Beijing Key Laboratory of Urban Oil and Gas Distribution Technology , China University of Petroleum, 102249 Beijing, China
e-mail: jiandu1997@163.com

J. Zheng (✉)
China Petroleum Planning and Engineering Institute, 100083 Beijing, China
e-mail: zhengjianqin_cup@163.com

H. Su et al. (eds.), *Advanced Intelligent Pipeline Management Technology*,
https://doi.org/10.1007/978-981-19-9899-7_11

At present, the methods of pipeline leakage detection and localization can be mainly divided into four types, negative pressure wave (NPW), transient inverse analysis (ITA), intelligent algorithms-based method, and data-driven methods. Despite some drawbacks, the above-mentioned methods have been widely used. When the pipeline leakage happens, these leakage detection methods attempt to determine the pipeline operation status based on hydrothermal process in the pipeline. For the study of pipeline operation condition, one of the important parts is the hydraulic calculation [3], but nonlinearity brings difficulties to hydraulic calculation [4].

For NPW, the calculated accuracy and sensitivity are higher and the lost is lower, and it has been involved in many practical projects [5]. During pipeline leakage, the relevant gradient and time information are collected by pressure signal collectors at both ends of the pipeline. The leak location can be determined by using signal processing methods. Despite the popularity of NPW in detecting pipeline leakage incidents, there are still some technical problems in the meantime. Initially, due to the transient disturbances in the system, there may generate false alarms results from similar negative pressure waves. Secondly, the method is insensitive in some leak scenarios, such as small leaks in pipelines. Some researchers have made progress in the improved NPW for pipeline leak detection and localization. Lu et al. [6] denoised the pipeline pressure signal by developing a small noise reduction method based on empirical mode decomposition. The proposed method is combined with NPW to detect pipeline leakage. Li et al. [7] simplified the methods of accurately determining NPW velocity transit time and interference based on liquid flow velocity attenuated by NPW.

ITA is proposed by Pudar and Liggett [8], and many researchers propose the combination of ITA and other algorithms to expand applicability of ITA. Vítkovský et al. [9] proposed that the uniqueness and quality of ITA solutions can be improved by the rapid input transients including maximum system response information through experimental observations. However, many restrictive conditions are required before using this method, such as fluctuations in leakage flow rate or artificial changes in valve operating conditions.

With the development of computing power and data storage capacity, the intelligent algorithms and data-driven methods are utilized for pipeline leakage detection. Zhang et al. [10] proposed a novel method for leak detection and localization in liquid pipelines by combining inverse hydrothermal transient analysis and improved particle swarm optimization (PSO). Huang et al. [11] developed a numerical method based on a combination of transient flow simulation and simulated annealing to detect leakage in pipeline networks. The intelligent algorithm and machine learning also provide contributions to many industries. Li et al. [12] proposed a leakage detection method by detecting transient in the pipeline, based on the nonlinear time series. Kang et al. [13] developed a novel model for leakage detection by combining a 1-D convolutional neural network with a support vector regression, then the leakage location is determined by applying localization algorithm based on graph theory. Fukuda and Mitsuoka developed a hybrid method for detecting smaller-scale leaks in pipeline; the statistical analysis technology and pressure gradient method were employed.

In recent years, there exist some studies for pipeline leakage. Liu et al. [14] proposed a comprehensive model, which contains a large leakage dynamic monitoring module and a small leakage static testing module. Xu et al. [15] analyzed the special characteristics near the leak source for small leaks; the fluid dynamic and aeroacoustics mathematical model were used. Liu et al. [16] expressed the original pressure data applying Markov chain by extracting the dynamic characteristics of the original pressure data and utilized switching rules to select different time scale models to accurately recognized pipeline states.

Although there is great progress made in the above-mentioned methods, the limitations can be summarized as follows:

(1) The existing methods are limited to determining only one possible combination of leakage parameters. In theory, upstream and downstream leaks and pressure combinations can also make the two measured parameters different. Therefore, a leakage accident may have different combinations of leakage parameters.
(2) Owing to the complexity of pipeline leaks, the present methods for leak detection commonly used are based on many assumptions; that is, the magnitude of the NPW propagation velocity is constant and so forth. However, due to complex physical mechanisms, these assumptions may against the actual operating conditions sometimes, which reduce the accuracy of leak detection consequently.

In recent years, a novel generative model, named generative adversarial networks (GANs), has been proposed [17]. From the point of view of fitting probability distributions, it can overcome both of these limitations. GANs aim to generate similar data to input data. Due to their excellent modeling abilities, GANs can adapt to complex data even with some implicit distributions and the data do not need to conform any specific assumptions. There has been a successful application for GAN in many frontier research areas, such as image processing [18] and natural language processing [19]. For this reason, this chapter proposes a GANs framework to detect pipeline leakage based on data-driven. As shown in Fig. 11.1, considering that the on-site leakage accidents barely happen and the statistical data of on-site pipeline leakage are lacked, the experimental simulation is carried out to generate leakage data. The measured data in SCADA system and the simulated data are input to the GANs framework, then the combinations of pipeline leakage parameters are generated. The generated data are applied to determine the leakage condition, providing a decision basis for on-site emergency repair.

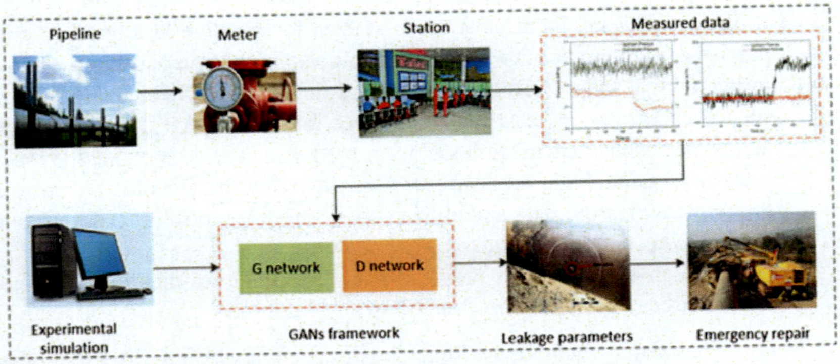

Fig. 11.1 Main process of the pipeline leakage detection method in this chapter

11.2 Intelligent Algorithm for Pipeline Leakage Detection

11.2.1 Description of the Generative Adversarial Networks

The GAN, which is proposed in 2014 [17], is an essentially generative framework. In general, GAN is applied to approximate the data distribution of the training set and then generate similar results with the same distribution. The traditional GAN includes two modules, a generator that matches the potential distribution in the training data and a discriminator that designed to distinguish the original training set samples from the synthetic ones.

The basic framework of GAN is depicted in Fig. 11.2. For GAN, two different neural networks try to against each other until the convergence condition is satisfied. The operation principle of GAN is that two neural networks are applied to play against each other when the Nash equilibrium is reached. More specifically, the GAN framework is composed of two functional parts, namely the generative (G) network and a discriminative (D) network. The G network is designed to approximate the distribution of the sample data and generates samples that are similar to the real training data, where the noise z follows a distribution (uniform, Gaussian, etc.). The G network aims to map the noise z to data space as $G\,(z, \theta G)$, where θG consists of the parameters for the G network, and $G(\bullet)$ is a multilayer neural network. For the (D) network, it is a two-class classifier employed to estimate the probability that a sample is derived from the trained data rather than the generated data. The D network will output a large probability when the sample is from real training data. If the sample is not from real training data, the D network obtains a small value. The D network mainly intends to maximize the classification accuracy, and the G network aims to maximize the probability that the data generated by the D network will be regarded as the real data.

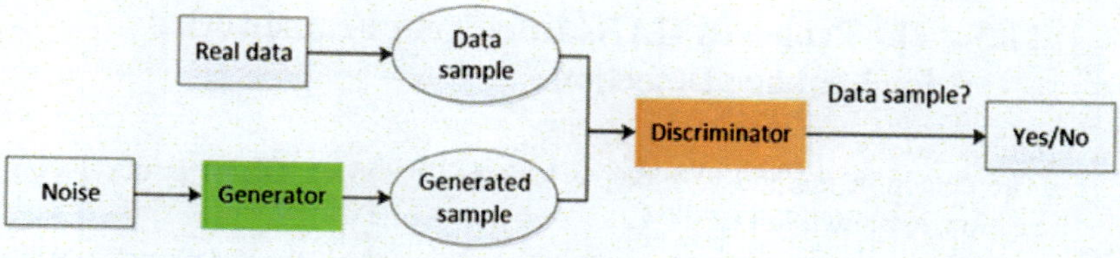

Fig. 11.2 Basic framework of the GAN

11.2.2 The Variants GANs

As research continues, GANs have also shown their shortcomings, such as model collapse and imbalanced training between generator and discriminator [20]. In recent years, great efforts have been made to improve the performance of GANs through various changes to standard GANs. In this chapter, according to different optimization methods, various GANs are divided into two categories, called structure-focusing and loss-focusing improvements, as shown in Fig. 11.3.

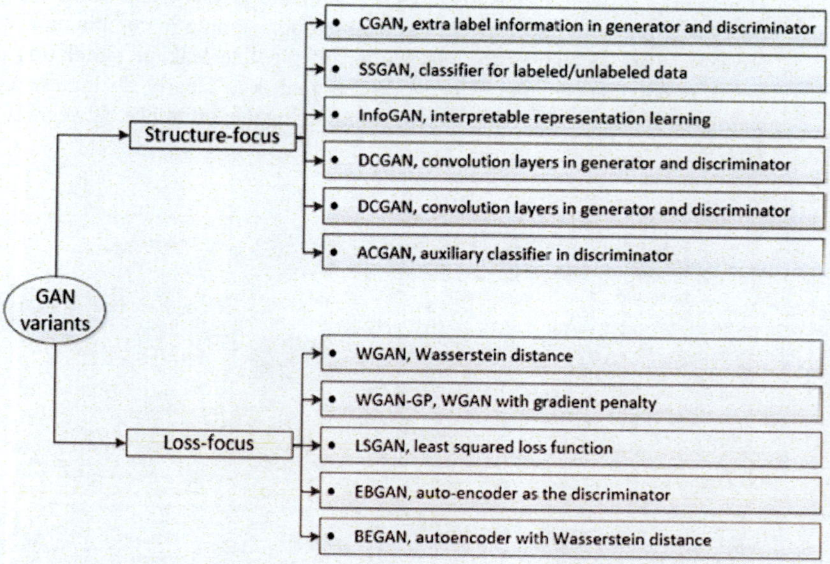

Fig. 11.3 Variants GANs

11.3 The Proposed GANs Intelligent Framework for Leakage Detection

The detection method of pipeline leakages based on GANs is developed in this section, as shown in Fig. 11.4.

The simulation method is applied to generate leakage data, and sample 1 is collected from simulated data. For G network, the inputs conform to specific distribution, and the generated data contain three leakage parameters, namely leakage time, coefficient, and location. Sample 2 contains measured data and the generated data. Measured data are collected employing the flow rate meters and pressure meters. The complete leak data are the combination of generated data and measured data and can be represented as $G(z) + x_{measured}$. Therefore, the GANs framework can be used to approximate the distribution of leakage data, regardless of whether the leakage data conform to the Gaussian distribution. The inputs of D network are two different samples, A and B. The D network outputs a probability value used to distinguish A from B.

The G and D networks update the network parameters with the help of the loss function. Training a GANs framework is closely related to the goals of the G and D networks as well as the loss function. For the GANs framework, the goal of the G network is to generate data as a generation sample in combination with the measured data to statisfy the distribution of pipeline leakage simulation data as much as possible. For the D network, the goal is to clearly distinguish between simulated and generated data, thereby maximizing the difference between $D(x)$ and $D(G(z) + x_{\text{measured}})$.

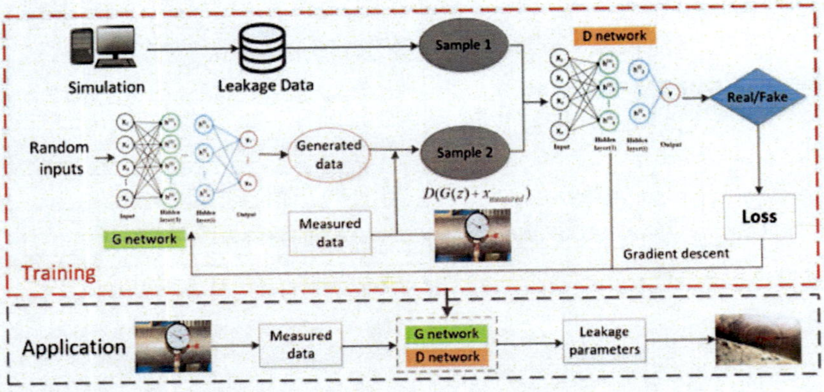

Fig. 11.4 Proposed detection method of pipeline leakage

11.4 Pipeline Leakage Experiments

11.4.1 Leakage Data Description

The experimental data are from a previous work [10]. Extensive pipeline leak data were obtained using relevant mathematical analysis. To illustrate the generality of the GAN framework for liquid pipelines, this section uses a hydraulic head to characterize the liquid pressure, as shown in Eq. (11.1).

$$H = \frac{P}{\rho g} \tag{11.1}$$

where H denotes the hydraulic head, P is the pipeline pressure, ρ is the liquid density, and g represents gravity acceleration.

The pipeline upstream head, down flow rate leakage time, coefficient, and location are initialized after obtaining operation parameters. After transient process analysis, the corresponding downstream head and upstream flow can be obtained.

It is significant to conduct data normalization, which can improve model accuracy and generlization ability by accelerating the convergence speed. In this chapter, the min–max method is applied to normalize each dimension of the datasets to values within the range of [0, 1] as follows:

$$\hat{x}_i = \frac{x_i - x_{\min}}{x_{\max} - x_{\min}} \tag{11.2}$$

where $\hat{x}_i$ is the normalized values of datasets and x_i is the original value. The $x_{\min}$ and $x_{\max}$ are the minimum value and maximum value of the datasets, respectively. After the data are normalized, the input data can be obtained.

11.4.2 Evaluation Metrics

The proposed model can generate various prediction results of pipeline leakage. Consequently, it is not appropriate to directly compare the prediction results of pipeline leakage with the real parameters. A precise detection model of pipeline leakage should be able to generate leakage data, which contains similar properties to the simulated data. Q_{up}, Q_{down}, P_{up}, and P_{down} are utilized to represent the upstream flow, downstream flow, upstream head, and downstream head. Then, the new upstream flow and downstream head can be acquired through taking the three leakage parameters, Q_{down} and P_{up} as the input of transient analysis process. Therefore, an error function is introduced to represent the performance of the proposed method for pipeline leakage modeling, as shown in Eq. (11.3).

$$\text{Error}(Q_{\text{up}}, \hat{Q}_{\text{up}}, P_{\text{down}}, \hat{P}_{down}) = \frac{1}{2}\frac{\left|Q_{\text{up}} - \hat{Q}_{\text{up}}\right|}{Q_{\text{up}}} + \frac{1}{2}\frac{\left|P_{\text{down}} - \hat{P}_{\text{down}}\right|}{P_{\text{down}}} \tag{11.3}$$

where Q_{up} and P_{down} are new upstream flow and downstream head obtained by transient analysis process.

11.4.3 Pipeline Leakage Cases

For the performance validation of the proposed pipeline leakage modeling method, the training principle of the GAN framework should first be determined. Through trial and error, the appropriate training algorithm was determined to be RMSprop, an adaptive learning rate method that overcomes the reduced learning rate problem.

After acquiring the GANs framework, two real-world pipeline leakages are taken as examples to verify the efficiency of the framework. The measurement data of the two examples are shown in Figs. 11.5 and 11.6. The actual leakage parameters are shown in Table 11.1. The measurement data are used as input, and the output is the generated (predicted) leakage parameters. After training, estimates of pipeline leakage parameters are obtained. To explain the accuracy and efficiency of the proposed method, 200 sets of generated parameters are randomly selected to form a set, and the error probability distributions are calculated.

It is demonstrated in example 1 and example 2 that, given measurement data, it is possible to estimate pipeline leak parameters using a trained GAN framework. Figure 11.6a shows that the downstream head drops sharply after 170 s, while the upstream head is faster than 170 s, indicating that the leakage time is less than 150 s and the leakage location is closer to the upstream, based on the negative pressure wave propagation. These similar results can be obtained in Figs. 11.5 and 11.6.

The effectiveness of this method is verified by taking the improved PSO [10] proposed in previous work and artificial neural network as comparisons. The leakage parameters for sample 1 and 2 are estimated by the improved PSO method and artificial neural network. A comparison of various methods of estimating pipeline leak parameters is presented in Table 11.1. Similarly, the highest probability values of the leak parameters generated by the GAN framework are also shown in Table 11.1. As shown in Table 11.1, the leakage locations of the improved PSO (2105.4 m) and the GANs framework (2100.0 m) are close to the actual location (2082.4 m). Furthermore, for example 1 or example 2, the leakage parameter prediction result of the GAN model is the best followed by the improved PSO, while the leakage parameter prediction result of the ANN is the worst. The errors described in Sect. 11.4.2 for these three models are calculated and depicted in Table 11.1 to further demonstrate the accuracy of the GANs model. For example 1, the estimated errors of the ANN improved PSO, and GANs frameworks are 8.4%, 4.8%, and 3.9%, respectively. For example 2, the errors are 4.6% (improved PSO), 7.3% (ANN), and 3.5%

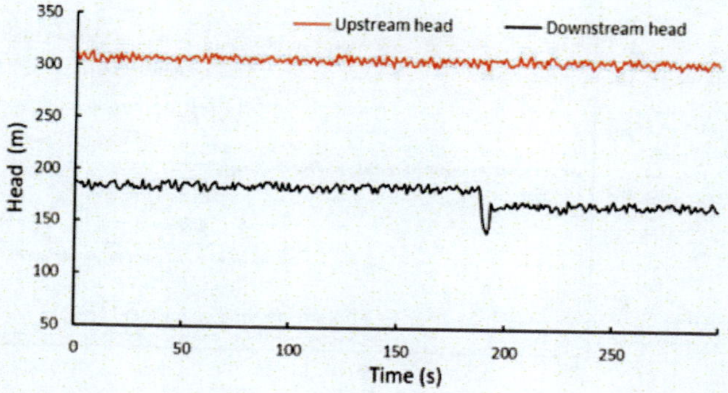

(a)

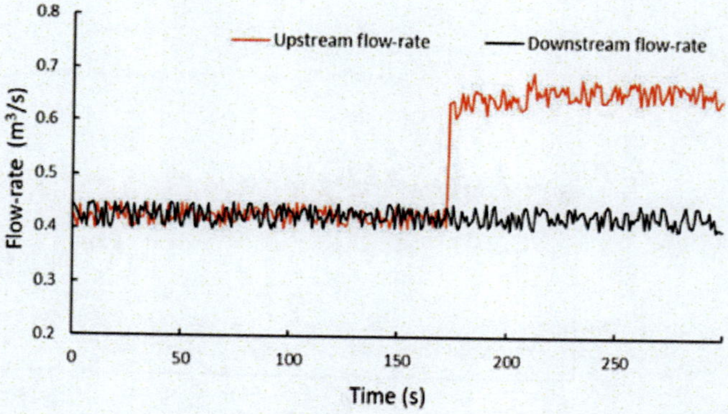

(b)

Fig. 11.5 Measured data of sample 1 (**a** head and **b** flow rate)

(GANs frameworks), respectively. By comparing the errors shown in Table 11.1, the efficiency of the GANs framework proposed in this work is verified.

11.5 Experimental Conclusion of Pipeline Leakage Detection

This chapter proposes a novel method based on GANs framework to estimate the leakage parameters accurately. The GAN is a generative framework and is made up with two modules (a generator and a discriminator). Typically, a GAN is applied to approximate the distribution of the data in the training set and then produce similar results with the same distribution. In this chapter, after obtaining the on-site flow rate and pressure of upstream and downstream measured data, the GANs framework is

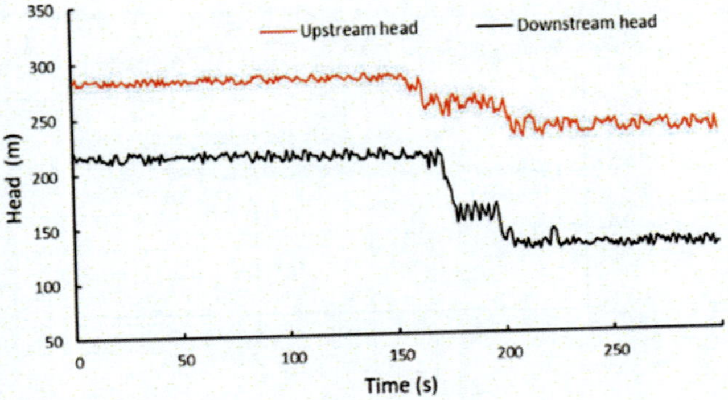

(a)

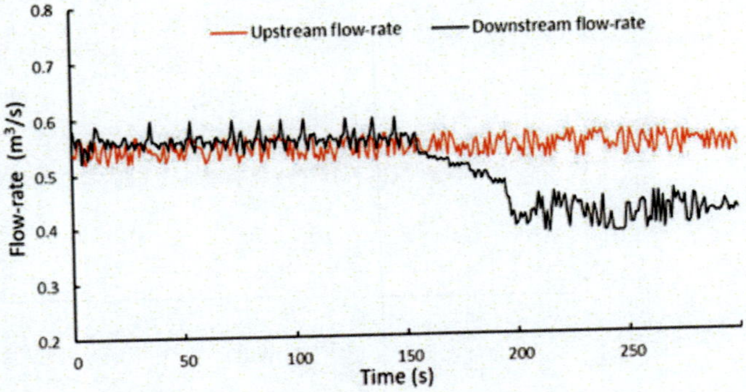

(b)

Fig. 11.6 Measured data of sample 2 (**a** head and **b** flow rate)

Table 11.1 Comparisons of different methods on pipeline leakage parameters estimation

		Leakage location (m)	Leakage coefficient	Leakage time (s)	Error (%)
Real	Example 1	2082.4	0.00320	141.4	/
	Example 2	5934.6	0.00212	155.8	/
Improved PSO	Example 1	2105.4	0.00327	138.9	4.8
	Example 2	5874.5	0.00205	153.6	4.6
ANN	Example 1	1964.3	0.00301	124.7	8.5
	Example 2	5788.9	0.00233	149.2	7.3
The GANs framework	Example 1	2100.0	0.00314	142.5	**3.9**
	Example 2	6000.0	0.00216	156.6	**3.5**

used to generate the potential leakage parameters combination, which contains coefficient leakage, time, and location. Two real-world pipeline leakages are applied to verify the accuracy and efficiency of the proposed framework. Results suggest that the proposed GANs framework outperforms in pipeline leakage parameters estimation and has a potential application value to guide on-site operation management.

References

1. Wang B, Liang Y, Zheng T, Yuan M, Zhang H. Optimisation of a downstream oil supply chain with new pipeline route planning. Chemical Engineering Research and Design. 2019;145:300–13. https://doi.org/10.1016/j.cherd.2019.03.009.
2. Black P. A review of pipeline leak detection technology. Pipeline systems. 1992:287–98. https://doi.org/10.1007/978-94-017-2677-1_23.
3. Liao Q, Liang Y, Xu N, Zhang H, Wang J, Zhou X. An MILP approach for detailed scheduling of multi-product pipeline in pressure control model. Chemical Engineering Research and Design. 2018;136:620–37. https://doi.org/10.1016/j.cherd.2018.06.016.
4. Sun J, Feng X, Wang Y, Deng C, Chu KH. Pump network optimization for a cooling water system. Energy. 2014;67:506–12. https://doi.org/10.1016/j.energy.2014.01.028.
5. Liu P, Zhou T, Li J. Application of negative pressure wave method in nuclear pipeline leakage detection. 2011 Asia-Pacific Power and Energy Engineering Conference: IEEE; 2011. p. 1–4. https://doi.org/10.1109/APPEEC.2011.5748950.
6. Lu W, Liang W, Zhang L, Liu W. A novel noise reduction method applied in negative pressure wave for pipeline leakage localization. Process Safety and Environmental Protection. 2016;104:142–9. https://doi.org/10.1016/j.psep.2016.08.014.
7. Li J, Zheng Q, Qian Z, Yang X. A novel location algorithm for pipeline leakage based on the attenuation of negative pressure wave. Process Safety and Environmental Protection. 2019;123:309–16. https://doi.org/10.1016/j.psep.2019.01.010.
8. Pudar RS, Liggett JA. Leaks in pipe networks. Journal of Hydraulic Engineering. 1992;118:1031–46. https://doi.org/10.1061/(ASCE)0733-9429(1992)118:7(1031).
9. Vítkovský JP, Lambert MF, Simpson AR, Liggett JA. Experimental observation and analysis of inverse transients for pipeline leak detection. Journal of Water Resources Planning and Management. 2007;133:519–30. https://doi.org/10.1061/(ASCE)0733-9496(2007)133:6(519).
10. Zhang H, Liang Y, Zhang W, Xu N, Guo Z, Wu G. Improved PSO-based method for leak detection and localization in liquid pipelines. IEEE Transactions on Industrial Informatics. 2018;14:3143–54. https://doi.org/10.1109/TII.2018.2794987.
11. Huang Y-C, Lin C-C, Yeh H-D. An optimization approach to leak detection in pipe networks using simulated annealing. Water Resources Management. 2015;29:4185–201. https://doi.org/10.1007/s11269-015-1053-4.
12. Li J, Liu W, Sun Z, Cui L. A new failure detection method and its application in leak monitor of pipeline. 2008 10th International Conference on Control, Automation, Robotics and Vision: IEEE; 2008. p. 1178–82. https://doi.org/10.1109/ICARCV.2008.4795688.
13. Kang J, Park Y-J, Lee J, Wang S-H, Eom D-S. Novel leakage detection by ensemble CNN-SVM and graph-based localization in water distribution systems. IEEE Transactions on Industrial Electronics. 2017;65:4279–89. https://doi.org/10.1109/TIE.2017.2764861.
14. Liu C, Li Y, Xu M. An integrated detection and location model for leakages in liquid pipelines. Journal of Petroleum Science and Engineering. 2019;175:852–67. https://doi.org/10.1016/j.petrol.2018.12.078.
15. Xu T, Chen S, Guo S, Huang X, Li J, Zeng Z. A small leakage detection approach for oil pipeline using an inner spherical ball. Process Safety and Environmental Protection. 2019;124:279–89. https://doi.org/10.1016/j.psep.2018.11.009.

16. Liu J, Zang D, Liu C, Ma Y, Fu M. A leak detection method for oil pipeline based on markov feature and two-stage decision scheme. Measurement. 2019;138:433–45. https://doi.org/10.1016/j.measurement.2019.01.029.
17. Goodfellow I, Pouget-Abadie J, Mirza M, Xu B, Warde-Farley D, Ozair S, et al. Generative adversarial nets. Advances in neural information processing systems. 2014;27.
18. Chen X, Duan Y, Houthooft R, Schulman J, Sutskever I, Abbeel P. Infogan: Interpretable representation learning by information maximizing generative adversarial nets. Advances in neural information processing systems. 2016;29.
19. Yu L, Zhang W, Wang J, Yu Y. Seqgan: Sequence generative adversarial nets with policy gradient. Proceedings of the AAAI conference on artificial intelligence 2017. https://doi.org/10.1609/aaai.v31i1.10804.
20. Mao X, Li Q, Xie H, Lau RY, Wang Z, Paul Smolley S. Least squares generative adversarial networks. Proceedings of the IEEE international conference on computer vision2017. p. 2794–802. https://doi.org/10.48550/arXiv.1611.04076

Chapter 12
Smart Emergency Management of Pipeline System

Weilong Ni and Zhengbing Li

Abstract Oil and gas resources are important strategic materials to ensure national economic, political, and military security. In emergency management, the supply of oil and gas can maintain the normal operation of various vehicles. When disasters cause supply disruptions, policymakers need to formulate emergency supply plans based on factors such as traffic impacts and resource demands caused by disasters, and deliver supply resources safely to disaster-stricken areas for replenishment within a specified time. This chapter summarizes the contents of emergency planning, such as site selection, resource allocation, and transportation mode, puts forward the mathematical model of related problems, and illustrates with two cases.

12.1 Emergency Management

The petrochemical industry occupies an important position in the national economy of various countries and is the basic industry and pillar industry of many countries [1]. Ensuring the safety and stability of the supply chain of petrochemical products is not only related to the development of a country's transportation industry, but also of great significance to the security of the country. In the past few years, various disasters have occurred frequently. These disasters cause heavy casualties and economic losses, threatening the development of the country and society [2]. After a disaster occurs, various emergency materials are needed to ensure the rescue work in the disaster area. Among them, oil and natural gas provide energy for various rescue transportation equipment and are one of the important materials to ensure the success of the rescue. Therefore, it is necessary to ensure the safety and stability of the supply of petroleum products and natural gas in order to ensure the effective distribution of disaster relief and emergency supplies.

W. Ni (✉) · Z. Li
Beijing Key Laboratory of Urban Oil and Gas Distribution Technology, China University of Petroleum, Fuxue Road No.18, Changping, Beijing 102249, China
e-mail: nwl_1997@163.com

Z. Li
e-mail: cup_lzb@163.com

H. Su et al. (eds.), *Advanced Intelligent Pipeline Management Technology*,
https://doi.org/10.1007/978-981-19-9899-7_12

For petroleum products, the emergency dispatch work is time-critical, and the demand in the supply process is uncertain [3]. Unlike other emergency supplies, petroleum products have many different modes of transportation, and pipelines are the most commonly used one [4]. At the same time, considering the volatility and flammability of petroleum products, stricter requirements for transportation carriers and storage facilities are required. In the rescue process, unlike other emergency supplies that can be stored in temporary warehouses, petroleum products can only be stored in existing oil depots near the disaster area. Therefore, it is necessary for decision makers to formulate an emergency dispatch plan for petroleum products according to the particularity of petroleum products and select different transportation modes to ensure the safety of the disaster relief process.

12.2 Emergency Scheduling

When disasters cause supply interruptions, decision makers are required to formulate emergency product supply plans according to the damage caused by disasters, and deliver supply resources safely to disaster-stricken areas within a specified time for supplementation. This chapter introduces the emergency dispatch system of petroleum products, including the transportation plan of emergency products and the repair plan of damaged roads.

This chapter is based on the analysis of the oil downstream supply chain system in Fig. 12.1. The emergency dispatch system consists of the refineries, the supply depots, the disaster-stricken cities, and the means of transportation connecting the above elements. Refineries and supply depots can transport oil to disaster-stricken cities by means of transportation such as pipelines, rail, and road [5, 6]. However, due to the damage to various infrastructures caused by disasters, the road for transporting emergency products may be hindered, resulting in the inability of emergency products to be transported to the disaster area in time. Therefore, choosing a suitable mode of transportation becomes a challenge. When developing a transportation plan for petroleum products, it is necessary to identify the amount of transportation by different modes of transportation from refineries or supply depots to downstream affected cities. Generally speaking, it is easier to choose a shipping origin. However, the occurrence of accidents such as pipeline rupture will prolong the transportation time or make it impossible to transport. Therefore, these accidents caused the difficulty of emergency dispatch. Therefore, these accidents are a major challenge for the emergency dispatch of petroleum products. This chapter considers the possibility of transport mode failure due to accidental damage and the resilience of post-disaster repairs, improving the fidelity of the model.

For the application of natural gas supply chain, there are mainly differences in the mode of transportation, but the general modeling and solving ideas are the same.

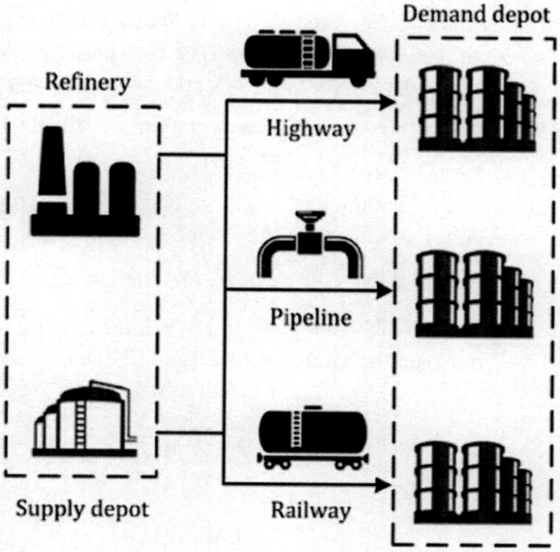

Fig. 12.1 Downstream oil supply chain system

12.3 Mathematical Method

12.3.1 Mathematical Model

In the emergency dispatch model established in this chapter, the uncertainty of the disaster area's demand for materials is considered. Since disasters may cause damage to roads and railways and rupture of pipelines, the model also takes into account the possibility that the mode of transportation may become unavailable, creating a multi-objective MILP model.

The mathematical model in this chapter includes the following five sets: the set of refinery and supply depot (I); the set of disaster node numbers (J); the set of transportation modes ($N = \{1, 2, 3\}$), where 1 represents pipeline transportation, 2 represents highway transportation, 3 represents railway transportation; the set of oil products ($O = \{1, 2\}$), where 1 denotes gasoline, 2 denotes diesel; the set of scheduling cycle (T), representing the number of days contained in one cycle.

The objective functions of the model are the minimum total cost (F_2) and total time (F_1).

$V_{t,n,i,j,o}$ is the variable representing the volume of transport by mode n from node i to node j in day t, and the parameter $q_{j,o}$ represents the delivery flow rate of oil at disaster node j. Thus, $V_{t,1,i,j,o}/q_{j,o}$ expresses the transportation time through pipeline. The binary variable $\mathrm{BW}_{t,n,i,j,o}$ indicates the choice of transportation modes, equaling to 0 if the mode n is not imposed. $th_{t,i,j}$ and $tr_{t,i,j}$ are parameters indicating the corrected. The binary variable $F_{t,n,i,j}$ represents whether the mode n needs to be fixed or not, $F_{t,n,i,j}= 1$ if the mode needs to be fixed. $tf_{n,i,j}$ is the maintenance time of mode n. The expression for the total time consists of two parts, as shown in Eq. (12.1). The first term $\left(\sum_{i\in I}\sum_{j\in J}\sum_{t\in T}\sum_{o\in O}\left(V_{t,1,i,j,o}/q_{j,o} + BW_{t,2,i,j,o}th_{t,i,j} + BW_{t,3,i,j,o}tr_{t,i,j}\right)\right)$ stands for transportation time, and the second one $\left(\sum_{i\in I}\sum_{j\in J}\sum_{t\in T}\sum_{n\in N} F_{t,n,i,j}tf_{n,i,j}\right)$ represents maintenance time. The parameters ct^u_n and cf^u_n represent the unit transportation cost and the unit maintenance cost. The transportation cost $\left(\sum_{t\in T}\sum_{i\in I}\sum_{j\in J}\sum_{o\in O}\left(p_{i,j}V_{t,1,i,j,o}lct^u_1 + l_{i,j}V_{t,2,i,j,o}ct^u_2 + lr_{i,j}V_{t,3,i,j,o}ct^u_3\right)\right)$ and maintenance cost $\left(\sum_{t\in T}\sum_{i\in I}\sum_{j\in J}\sum_{n\in N} F_{t,n,i,j}ld_{n,i,j}cf^u_n\right)$ of all modes make up the total cost (i.e., Eq. (12.2)).

$$\min F_1 = \sum_{i\in I}\sum_{j\in J}\left(\sum_{t\in T}\left(\sum_{o\in O}\left(\frac{V_{t,1,i,j,o}}{q_{j,o}} + \mathrm{BW}_{t,2,i,j,o}th_{t,i,j}\right.\right.\right.$$
$$\left.\left.\left. + \mathrm{BW}_{t,3,i,j,o}tr_{t,i,j}\right) + \sum_{n\in N} F_{t,n,i,j}tf_{n,i,j}\right)\right) \tag{12.1}$$

$$\min F_2 = \sum_{t\in T}\sum_{i\in I}\sum_{j\in J}\left(\sum_{o\in O}\left(lp_{i,j}V_{t,1,i,j,o}ct^u_1 + l_{i,j}V_{t,2,i,j,o}ct^u_2\right.\right.$$
$$\left.\left. + lr_{i,j}V_{t,3,i,j,o}ct^u_3\right) + \sum_{n\in N} F_{t,n,i,j}ld_{n,i,j}cf^u_n\right) \tag{12.2}$$

In terms of constraints, the following constraints are often considered: transportation constraints, time constraints, material balance constraints, inventory constraints, vehicle constraints, and the maintenance constraints for transportation modes.

12.3.2 Solving Method

For a multi-objective problem, the result of one objective may not satisfy the optimality of other objectives, so the solution result may not satisfy the situation that all objective functions are optimal solutions at the same time. A compromise between objectives is required to make each objective in the model as optimal as possible [7]. The multi-objective optimization solutions obtained in this way are a set of non-unique Pareto optimal solutions. Pareto optimality means that at least one objective function is improved without making any objective function worse. In order to generate a subset of Pareto sets, the augmented ε-constraint method (AUGMECON) [8] is used in this chapter. The other objective functions are converted into constraints in AUGMECON, so only one objective function will be optimized. The mathematical expression of AUGMECON is shown in Eq. (12.3).

$$
\begin{aligned}
&\min\left(f_1(x) + eps \times \left(s_2 + s_3 + \cdots + s_q\right)\right) \\
&st \\
&f_2(x) - s_2 = e_2, \\
&f_3(x) - s_3 = e_3, \\
&\ldots \\
&f_q(x) - s_q = e_q, \\
&x \in S \text{ and } s_i \in R^+
\end{aligned}
\tag{12.3}
$$

The objective functions in multi-objective programming are represented by $f_2(x)$, $f_3(x)$, ..., $f_q(x)$, respectively, the surplus variables s_2, s_3, ..., s_q introduce in the constraints, $e_2, e_3, \ldots, e_q$ are the relevant iteration parameters, and eps is a distribution smaller value between 10^{-3} and 10^{-6}.

This chapter proposes a bi-objective model, where $f_1(x)$ corresponds to Eq. (12.1) and $f_2(x)$ corresponds to Eq. (12.2). Firstly, by solving function $f_2(x)$ as a single-objective function, its range can be obtained. Then, the objective function $f_2(x)$ is divided into equal amounts to obtain q equidistant iteration parameters. Finally, the obtained parameters are substituted into the original bi-objective model to obtain the Pareto set.

12.3.3 Robust Optimization

This chapter adopts the robust optimization method to solve the data uncertainty problem. Compared with other methods, the robust optimization method only needs to search for the distribution that satisfies the conditions to find the optimal solution, thus avoiding the need to master the specific distribution of the uncertain set. This chapter uses the Γ-robustness approach to solve the mathematical model [9]. Assuming that the uncertain demand in this model fluctuates in an interval, by introducing disturbance parameters and adjusting control parameters according to uncertain factors, the optimality and robustness of the understanding are guaranteed.

When a linear model contains constraints with uncertain parameters, its mathematical expression is usually as follows:

$$
\sum_{p \in P_l} a_{lp} x_{lp} \le b_l \tag{12.4}
$$

x_{lp} represents the variable, the certain parameter is represented by b_l, a_{lp} is the uncertain parameter, and P_l represents the set of the uncertain parameters. Assuming that a_{lp} is a random number whose value range is between $\left[\bar{a_{lp}} - \tilde{a_{lp}}, \bar{a_{lp}} + \tilde{a_{lp}}\right]$, where $\tilde{a_{lp}}$ represents a maximum deviation from the mean value $\bar{a_{lp}}$.

In order to solve the model, the disturbance parameters and scale deviations $z_{lp} = (a_{lp} - \bar{a_{lp}})/\tilde{a_{lp}} \in [-1, 1]$ need to be defined first. This chapter assumes that the maximum deviation of the total scaled deviation is $\Gamma_l(\sum_{p \in P_l} |z_{lp}| \leq \Gamma_l)$. Γ_l is a parameter that varies from the range $[0, |P_l|]$. If $\Gamma_l = 0$, the robust model be a model without uncertainty, and if $\Gamma_l = |P_l|$, the solved target model is treated as an absolutely robust optimization model. Therefore, it is necessary to adjust the parameter Γ_l to control the disturbance range, and it should ensure that the value of Γ_l can conform to the characteristics of the initial stage of the emergency. The constraint (12.4) is transformed into a constraint (12.5) by introducing a protection function $\beta_l(x, \Gamma_l)$. Under the transformation of the above model, it can obtain the value of the protection function $\beta_l(x, \Gamma_l)$. Due to the effect of the nonlinearization of constraint (12.5), linearization is required using the strong duality theorem of SP1 duality.

$$\sum_p \bar{a_{lp}}\, x_{lp} + \beta_l(x, \Gamma_l) \leq b_l \tag{12.5}$$

$$\begin{aligned} SP1: \ & \beta_l(x, \Gamma_l) = \max \sum_{p \in P_l} a_{lp} |\tilde{x}_p| z_{lp} \\ s.t \ & \sum_{p \in P_l} z_{lp} \leq \Gamma_l \\ & 0 \leq z_{lp} \leq 1, \forall p \in P_l \end{aligned} \tag{12.6}$$

Using the above method, the original model is transformed into a robust model that takes into account demand uncertainty. Solving the model can get an oil emergency dispatch scheme considering demand uncertainty.

12.4 Case Study

12.4.1 Basic Data

Since the occurrence of earthquake disasters will have a serious impact on many areas, the facilities and traffic routes along the line will be damaged to varying degrees, resulting in an increase in the demand for emergency resources. Therefore, it is urgent to formulate a reasonable emergency dispatch plan for petroleum products to ensure the safety and stability of rescue operations.

This chapter takes the earthquake that occurred in Sichuan, China, in 2008 as a case background. Figure 12.2 shows the downstream oil supply chain system near the disaster area. There are 10 disaster areas (D1-D10) in the picture that need emergency supplies, which include different petroleum products (O1-O2). Petroleum products

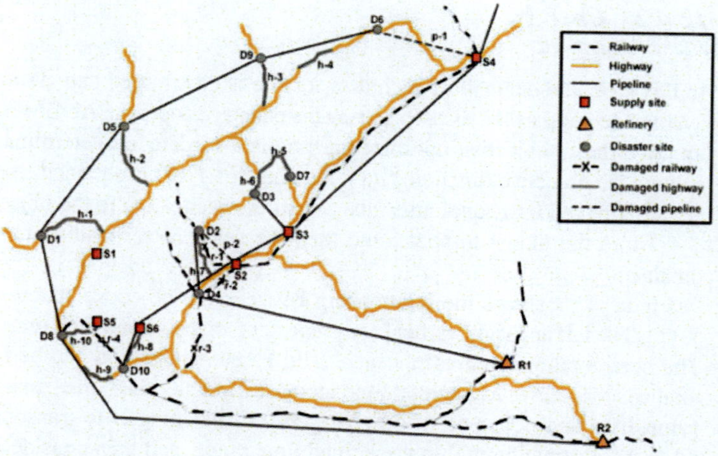

Fig. 12.2 Distribution map of supply and disaster points

can be supplied from six nearby supply depots (S1-S6), while two refineries (R1-R2) can also be responsible for supplying emergency petroleum products. There are three pipelines under this system. The first one goes from R1 to D2, via D4. The second pipeline runs from R2 to D9, via D8, D1, and D5. The last one is S4-S3-S2-S6-D10, in which there are three branches, S4-D6, S2-D2, and S3-D3. Two of the pipelines (S2-D2 and S4-D6) are damaged due to the disaster. The system scheduling period in this case is 6 days. Assuming that each refinery and supply depot can process sufficient oil products, the daily demand and consumption of each oil product in each disaster area is the same, and the scheduling time of each disaster area does not exceed 24 h.

Earthquakes may cause varying degrees of damage to various modes of transportation. This chapter divides the degree of earthquake damage to highways into three levels by the different parameter α and λ_2. When the damage degree $\alpha_{t,i,j}$ is less than α, it is assumed to be slightly damaged, and the oil can still be transported by highways. When the degree of damage is greater than α but less than λ_2, it is assumed to be severely damaged, and only when the road returns to its normal function can the highways be used for transportation. When the damage degree is greater than λ_2, it is considered that the highway cannot be repaired to the normal transportation state, and the transportation mode is considered to be completely destroyed. Similarly, the damage degree of the railway $\beta_{t,i,j}$ is also divided into three grades according to the above classification.

This chapter will analyze two types of cases in this region. Case 1 is a multi-objective planning problem with determination of the needs of the disaster area, solved by AUGMECON. Case 2 considers the uncertainty of demand in the disaster area on the basis of case 1 and adopts the robust optimization method to solve the model.

12.4.2 Case 1

In this case, the demand of each disaster site is determined and assumed to be the average demand of the disaster area. This chapter adopts the AUGMECON method to solve the bi-objective optimization problem of demand determination. By first solving the objective function $f_2(x)$, the range of $f_2(x)$ is obtained, and the range is equally divided into equal intervals. Substitute each value in the objective function $f_2(x)$ into the model to solve, and then the value corresponding to $f_1(x)$ can be obtained.

Figure 12.3 shows the relationship between the objective functions $f_1(x)$ and $f_2(x)$. The left and right vertical axes represent total cost and total time, respectively. The horizontal axis represents each valid Pareto solution. It can be found that the total cost decreases as the total time increases, and the final time approaches a fixed value. The first objective of oil emergency dispatching is to transport the needed oil to the destination within a specified time range, and then consider the total cost of the transportation process on this basis. It can be seen from the figure that the objective function $f_1(x)$ begins to converge at the 6th group of Pareto solutions. By comparison, it can be found that for the objective function $f_1(x)$ (the total time), the difference between the 6th group of solutions and the subsequent solutions is small, but the objective function $f_2(x)$ (the total cost) of the subsequent solutions is higher than that of the 6th group. Therefore, this chapter chooses the sixth group as the final plan of case 1.

As shown in Fig. 12.4, the scheduling scheme is different for each day in a cycle. Taking into account the cost gap between different transportation modes and the overall scheduling time, it can be found that when the disaster area is close to the supply depot and the demand is small, it is more economical to use road transportation products. On the contrary, when the disaster area is far away from the supply depot

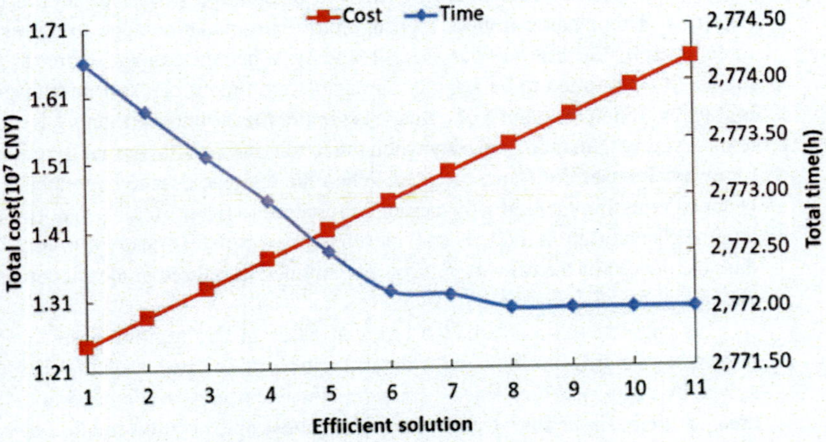

Fig. 12.3 Pareto solution distribution of f_1 and f_2

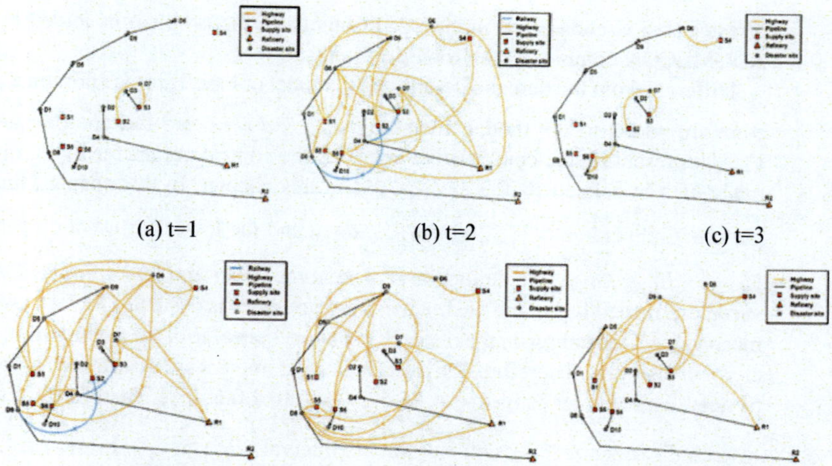

(a) t=1 (b) t=2 (c) t=3

(d) t=4 (e) t=5 (f) t=6

Fig. 12.4 Detailed scheme

and the demand is large, pipelines can be used in order to be more cost-effective. In addition, railways can be used as a way to supplement the needs of disaster areas, suitable for long-distance disaster areas. When $t = 1$, since most of the highway and railways are damaged, the supply point mainly transports petroleum products to the affected areas through pipelines. The pipelines and railways leading to D7 were damaged, so only oil products can be transported by road. When $t = 2$, the damaged road is repaired. For disaster areas such as D4, products can be supplied to disaster areas by more transportation modes such as road and rail, so as to reduce the transportation cost. The railway is mainly to supplement the remaining demand in the disaster area. The analysis of the scheduling plan for the remaining four days is the same as the above two days. Oil cannot be transported to these areas by road as some of the roads are completely destroyed and cannot be repaired within a dispatch cycle, the delivery of oil to these areas cannot be done by highways. For pipeline and railway, pipelines can transport more oil at a lower cost, so the pipeline is the first choice for the above disaster areas.

12.4.3 Case 2

In case 1, the complexity of the model is reduced by averaging the requirements over a period. But in reality, the demand for all kinds of oil in disaster areas is often uncertain. The actual situation is that when the demand deviates from the mean value, the solution result will be quite different from the optimal result, and the solution scheduling time or cost will increase. Therefore, the demand uncertainty in each disaster area is considered in case 2. In this case, the problem is solved by using AUGMECON combined with robust optimization.

Different from the demand deterministic model of case 1, some additional parameters are added to the model, including $\widetilde{dl}_{t,j,o}$, $\widetilde{du}_{t,j,o}$ and the control parameter $\Gamma_{t,j,o}$ representing the conservative level. As the control parameter $\Gamma_{t,j,o}$, the value range of it is defined $\{0, 0.1, 0.2, \ldots, 1\}$ in this chapter. In this chapter, the upper perturbation parameter $\gamma_{t,j,o} = \widetilde{du}_{t,j,o} / \overline{d}_{t,j,o}$ and the lower perturbation parameter $\gamma'_{t,j,o} = \widetilde{dl}_{t,j,o} / \overline{d}_{t,j,o}$ are defined to represent the upper and lower limits of demand variation, respectively, and each uncertain parameter is assumed to vary around its mean value. Depending on the research problem, the value of the perturbation parameter will also vary. Therefore, the disturbance parameters are determined empirically by need. This chapter defines $\gamma'_{t,j,o} = 0, \gamma_{t,j,o} \in \{0.2, 0.5, 0.8\}$. Therefore, the demand of each disaster site $d_{t,j,o}$ will vary on the interval $\left[\overline{d}_{t,j,o}, \overline{d}_{t,j,o}(1 + \gamma_{t,j,o})\right]$.

Compared with case 1, this case has undergone significant changes in the scheduling scheme. Figure 12.5 shows that Case 2 also has a different daily schedule over a period. When $t = 1$, pipelines are the main way of transporting oil from supply points to disaster areas. The only way to supply D7 is by road. In case 2, uncertainty in demand results in a change in the number of tankers sent from the supply point. In case 1, S3 sends 27 tankers to D7, while in case 2, S3 sends 25 tankers to D7. When $t = 2$, the damaged road will be repaired, and the disaster areas such as D4 can be supplied with oil by highway or railway. Due to uncertainty, the demand for petroleum products in the disaster areas in case 2 and case 1 is inconsistent. Although the scheduling scheme obtained in case 1 can be substituted into case 2, the total time and total cost of this scheme are not optimal results. Therefore, case 2 requires a scheduling scheme with less time and less transportation cost.

For example, from the results of case 1, it can be found that the D4 disaster area can receive gasoline from S3, S6, and R1, and the transportation modes include rail, road, and pipeline. But as the demand changes, only R1 supplies gasoline to D4 through the pipeline in case 2, resulting in a reduction in dispatch time. Similarly, when $t = 3$, S4 supplies gasoline to D6 by highway in case 1. While in case 2, R2 supplies gasoline to D6 by pipeline. The demand for gasoline in D6 has increased due to changes in demand in various disaster areas due to uncertainty. Due to the impact of demand uncertainty, the demand of D6 for gasoline increases. It can be found that the scheduling scheme given in case 2 not only reduces the scheduling time, but also reduces the transportation cost compared with the previous one.

In case 2, considering the uncertainty of the demand in the disaster area, the total supply of different transportation modes to the disaster area can still meet the demand in a period.

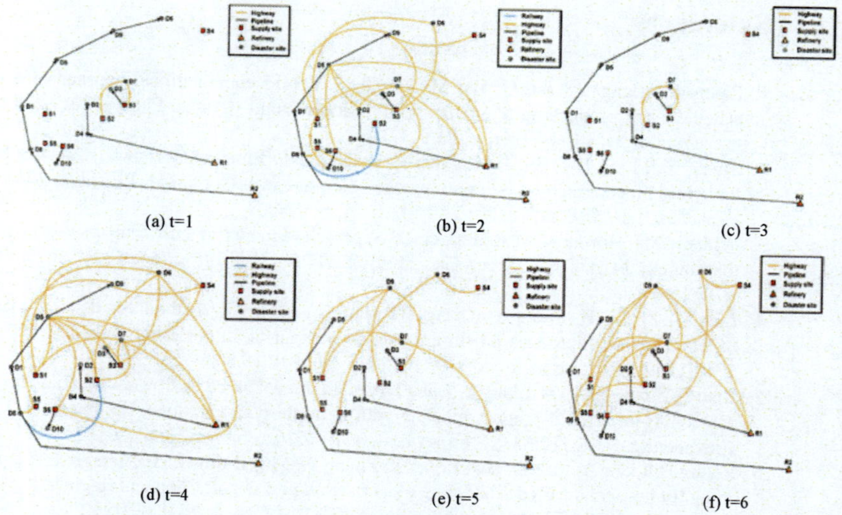

Fig. 12.5 Detailed scheme

12.5 Conclusion

Considering demand uncertainty and path effectiveness, a bi-objective MILP model for post-disaster emergency dispatch of petroleum products based on robust optimization is proposed with minimum total time and minimum cost as objective functions in this chapter. The time mainly considers the transportation time of the product and the maintenance time of the transportation method, and the cost includes the transportation cost of oil and the maintenance cost of the transportation method. AUGMECON combined with robust optimization method was used to solve the model.

This chapter presents two sets of cases to validate the proposed model. Under the conditions of demand certainty in case 1 and uncertainty in case 2, the solution is obtained, and a scheme that satisfies the scheduling period is obtained. In a multi-objective model, it is not possible to optimize all objective functions simultaneously. A reduction in total cost in this model results in an increase in total time. The model can provide effective decision-making basis for oil emergency dispatching under uncertain demand and greatly improve the robustness and economy of emergency dispatching. There are multiple modes of transportation between the supply point and the disaster area, and the choice of route will also affect the transportation plan.

References

1. Zhang H, Liang Y, Liao Q, Wu M, Yan X. A hybrid computational approach for detailed scheduling of products in a pipeline with multiple pump stations. Energy. 2017;119:612–28. https://doi.org/10.1016/j.energy.2016.11.027
2. Zheng Y-J, Ling H-F, Shi H-H, Chen H-S, Chen S-Y. Emergency railway wagon scheduling by hybrid biogeography-based optimization. Computers & Operations Research. 2014;43:1–8. https://doi.org/10.1016/j.cor.2013.09.002
3. Widener MJ, Horner MW. A hierarchical approach to modeling hurricane disaster relief goods distribution. Journal of Transport Geography. 2011;19:821–8. https://doi.org/10.1016/j.jtrangeo.2010.10.006
4. Liao Q, Zhang H, Wang Y, Zhang W, Liang Y. Heuristic method for detailed scheduling of branched multiproduct pipeline networks. Chemical Engineering Research and Design. 2018;140:82–101. https://doi.org/10.1016/j.cherd.2018.10.003
5. Liao Q, Zhang H, Xu N, Liang Y, Wang J. A MILP model based on flowrate database for detailed scheduling of a multi-product pipeline with multiple pump stations. Computers & Chemical Engineering. 2018;117:63–81. https://doi.org/10.1016/j.compchemeng.2018.05.002
6. Zhang H, Liang Y, Liao Q, Gao J, Yan X, Zhang W. Mixed-time mixed-integer linear programming for optimal detailed scheduling of a crude oil port depot. Chemical Engineering Research and Design. 2018;137:434–51. https://doi.org/10.1016/j.cherd.2018.07.013
7. Qiu R, Zhang H, Gao X, Zhou X, Guo Z, Liao Q, et al. A multi-scenario and multi-objective scheduling optimization model for liquefied light hydrocarbon pipeline system. Chemical Engineering Research and Design. 2019;141:566–79. https://doi.org/10.1016/j.cherd.2018.11.018
8. Mavrotas G. Effective implementation of the ε-constraint method in Multi-Objective Mathematical Programming problems. Applied Mathematics and Computation. 2009;213:455–65. https://doi.org/10.1016/j.amc.2009.03.037
9. Moradi S, MirHassani SA. Robust scheduling for multi-product pipelines under demand uncertainty. The International Journal of Advanced Manufacturing Technology. 2016;87:2541–9. https://doi.org/10.1007/s00170-016-8561-0